Marx Machiavelli Chesterton Cooper Emerson Joyce
Abbott Hardy Montaigne Haggard Austen
Defoe Melville Hugo
Christie Eliot Grimm
Stoker Carroll Molière
Wilde Maupassant Byron
Engels Schiller
Garnett Einstein Fitzgerald Smith
Goethe Hawthorne Kafka
Dostoyevsky Hall
Cotton Kipling Doyle
Baum Leslie Dumas Henry Willis
Nietzsche
Flaubert Turgenev Balzac
Stockton Vatsyayana Crane
Burroughs Whitman Verne
Curtis Tocqueville Gogol Vinci
Homer Widger Tolstoy Busch
Darwin Thoreau
Potter Freud Zola Twain Plato Scott
Lawrence
Kant Jowett Stevenson Dickens
Andersen Burton Hesse Harte
London Descartes Cervantes
Poe Aristotle Wells Voltaire
Hale James Hastings
Cooke
Bunner Shakespeare
Richter Cid Chambers Irving
Doré Swift Dante Chekhov da Shaw Wodehouse Benedict Alcott
Pushkin
Newton

tredition®

tredition was established in 2006 by Sandra Latusseck and Soenke Schulz. Based in Hamburg, Germany, tredition offers publishing solutions to authors and publishing houses, combined with worldwide distribution of printed and digital book content. tredition is uniquely positioned to enable authors and publishing houses to create books on their own terms and without conventional manufacturing risks.

For more information please visit: www.tredition.com

TREDITION CLASSICS

This book is part of the TREDITION CLASSICS series. The creators of this series are united by passion for literature and driven by the intention of making all public domain books available in printed format again - worldwide. Most TREDITION CLASSICS titles have been out of print and off the bookstore shelves for decades. At tredition we believe that a great book never goes out of style and that its value is eternal. Several mostly non-profit literature projects provide content to tredition. To support their good work, tredition donates a portion of the proceeds from each sold copy. As a reader of a TREDITION CLASSICS book, you support our mission to save many of the amazing works of world literature from oblivion. See all available books at www.tredition.com.

Project Gutenberg

The content for this book has been graciously provided by Project Gutenberg. Project Gutenberg is a non-profit organization founded by Michael Hart in 1971 at the University of Illinois. The mission of Project Gutenberg is simple: To encourage the creation and distribution of eBooks. Project Gutenberg is the first and largest collection of public domain eBooks.

Anecdotes of Dogs

Edward Jesse

Imprint

This book is part of TREDITION CLASSICS

Author: Edward Jesse
Cover design: Buchgut, Berlin – Germany

Publisher: tredition GmbH, Hamburg - Germany
ISBN: 978-3-8472-2489-1

www.tredition.com
www.tredition.de

ANECDOTES

OF

DOGS.

BY

EDWARD JESSE, Esq.

"Histories are more full of examples of the fidelity of dogs than of friends."—Pope.

With numerous Engravings.

LONDON:
HENRY G. BOHN, YORK STREET, COVENT GARDEN.
MDCCCLVIII.

LONDON:
Printed by G. Barclay, Castle St. Leicester Sq.

PREFACE.

The character, sensibilities, and intellectual faculties of animals have always been a favourite study, and they are, perhaps, more strongly developed in the dog than in any other quadruped, from the circumstance of his being the constant companion of man. I am aware how much has been written on this subject, but having accumulated many original and interesting anecdotes of this faithful animal, I have attempted to enlarge the general stock of information respecting it. It is a pleasing task, arising from the conviction that the more the character of the dog is known, the better his treatment is likely to be, and the stronger the sympathy excited in his behalf.

Let me hope, that the examples which are given in the following pages will help to produce this effect, and that a friend so faithful, a protector so disinterested and courageous, will meet with that kindness and affection he so well deserves.

It is now my grateful duty to express my thanks to those friends who have so kindly contributed original anecdotes to this work, and especially to Lady Morgan and Mrs. S. Carter Hall for their remarks on the Irish wolf-dog.

I have also to acknowledge my obligations for various anecdotes illustrative of the character of peculiar dogs, extracted from Colonel Hamilton Smith's volumes in the Naturalist's Library and Captain Brown's interesting sketches; as well to the Editor of the "Irish Penny Magazine" for his extremely well-written account of the Irish wolf-dog; and to other sources too numerous to mention.

The present new edition is considerably enlarged, both in matter and plates, and, to suit the taste of the age is presented in a cheap and popular form.

My Publisher has, as usual, lent his aid, and is responsible for some of the additional anecdotes, for the account of the *Setter*, and for all after page 458, including the chapter "On Feeding and Management."

Edward Jesse.

East Sheen, Sept. 1858.

SPANIEL AND NEWFOUNDLAND DOGS.

A French writer has boldly affirmed, that with the exception of women there is nothing on earth so agreeable, or so necessary to the comfort of man, as the dog. This assertion may readily be disputed, but still it will be allowed that man, deprived of the com Page 2 panionship and services of the dog, would be a solitary and, in many respects, a helpless being. Let us look at the shepherd, as the evening closes in and his flock is dispersed over the almost inaccessible heights of mountains; they are speedily collected by his indefatigable dog—nor do his services end here: he guards either the flock or his master's cottage by night, and a slight caress, and the coarsest food, satisfy him for all his trouble. The dog performs the services of a horse in the more northern regions; while in Cuba and some other hot countries, he has been the scourge and terror of the runaway negroes. In the destruction of wild beasts, or the less dangerous stag, or in attacking the bull, the dog has proved himself to possess pre-eminent courage. In many instances he has died in the defence of his master. He has saved him from drowning, warned him of approaching danger, served him faithfully in poverty and distress, and if deprived of sight has gently led him about. When spoken to, he tries to hold conversation with him by the movement of his tail or the expression of his eyes. If his master wants amusement in the field or wood, he is delighted to have an opportunity of procuring it for him; if he finds himself in solitude, his dog will be a cheerful and agreeable companion, and maybe, when death comes, the last to forsake the grave of his beloved master.

There are a thousand little facts connected with dogs, which many, who do not love them as much as I Page 3 do, may not have observed, but which all tend to develope their character. For instance, every one knows the fondness of dogs for warmth, and that they never appear more contented than when reposing on the rug before a good fire. If, however, I quit the room, my dog leaves his warm berth, and places himself at the door, where he can the better hear my footsteps, and be ready to greet me when I re-enter. If I am preparing to take a walk, my dog is instantly aware of my intention. He frisks and jumps about, and is all eagerness to accompany me. If I am thoughtful or melancholy, he appears to sympathise with me; and, on the contrary, when I am disposed to be merry, he shows by his manner that he rejoices with me. I have often watched the effect which a change in my countenance would produce. If I frown or look severe, but without saying a word or uttering a sound, the effect is instantly seen by the ears dropping, and the eyes showing unhappiness, together with a doubtful movement of the tail. If I afterwards smile and look pleased, the tail wags joyously, the eyes are filled with delight, and the ears even are expressive of happiness. Before a dog, however, arrives at this knowledge of the human countenance, he must be the companion of your walks, repose at your feet, and receive his food from your hands: treated in this manner, the attachment of the dog is unbounded; he becomes fond, intelligent, and grateful. Whenever Stanislas, the unfortunate King of Poland, wrote to his daughter, he always concluded Page 4 his letter with these words—"Tristan, my companion in misfortune, licks your feet:" thus showing that he had still one friend who stuck to him in his adversity. Such is the animal whose propensities, instincts, and habits, I propose to illustrate by various anecdotes.

The propensities of the dog, and some of them are most extraordinary, appear to be independent of that instinct which Paley calls, "a propensity previous to experience, and independent of instruction." Some of these are hereditary, or derived from the habits of the parents, and are suited to the purposes to which each breed has long been and is still applied. In fact, their organs have a fitness or unfitness for certain functions without education;—for instance, a very young puppy of the St. Bernard breed of dogs, when taken on snow for the first time, will begin to scratch it with considerable eagerness. I have seen a young pointer of three or four weeks old stand steadily on first seeing poultry, and a well-bred terrier puppy will show a great deal of ferocity at the sight of a rat or mouse.

Sir John Sebright, perhaps the best authority that can be quoted on this subject, says that he had a puppy of the wild breed of Australia; that the mother was with young when caught, and the puppy was born in the ship that brought her over. This animal was so like a wolf, not only in its appearance, but in all its habits, that Sir John at first doubted if it really were a dog, but this was afterwards proved by experiment.

Of all the propensities of the brute creation, the well- Page 5 known attachment of the dog to man is the most remarkable, arising probably from his having been for so many years his constant companion, and the object of his care. That this propensity is not instinctive is proved, by its not having existed, even in the slightest degree, in the Australian dog.

Sir John Sebright kept this animal for about a year, almost always in his room. He fed him himself, and took every means that he could think of to reclaim him, but with no effect. He was insensible to caresses, and never appeared to distinguish Sir John from any other person. The dog would never follow him, even from one room to another; nor would he come when called, unless tempted by the offer of food. Wolves and foxes have shown much more sociability than he did. He appeared to be in good spirits, but always kept aloof from the other dogs. He was what would be called tame for an animal in a menagerie; that is, he was not shy, but would allow strangers to handle him, and never attempted to bite. If he were led near sheep or poultry, he became quite furious from his desire to attack them.

Here, then, we see that the propensities that are the most marked, and the most constant in every breed of domestic dogs, are not to be found in animals of the same species in their natural state, or even in their young, although subjected to the same treatment from the moment of their birth.

Notwithstanding the above-mentioned fact, we may, Page 6 I think, consider the domestic dog as an animal *per se*; that is, that it neither owes its origin to the fox nor wolf, but is sprung from the wild dog. In giving this opinion, I am aware that some naturalists have endeavoured to trace the origin of the dog from the fox; while others, and some of the most eminent ones, are of opinion that it sprung from the wolf. I shall be able to show that the former is out of the question. The wolf, perhaps, has some claim to be considered as the parent animal, and that he is susceptible of as strong attachment as the dog is proved by the following anecdote, related by Cuvier.

He informs us, that a young wolf was brought up as a dog, became familiar with every person whom he was in the habit of seeing, and in particular, followed his master everywhere, evincing evident chagrin at his absence, obeying his voice, and showing a degree of submission scarcely differing in any respect from that of the domesticated dog. His master, being obliged to be absent for a time, presented his pet to the Ménagerie du Roi, where the animal, confined in a den, continued disconsolate, and would scarcely eat his food. At length, however, his health returned, he became attached to his keepers, and appeared to have forgotten all his former affection; when, after an absence of eighteen months, his master returned. At the first word he uttered, the wolf, who had not perceived him amongst the crowd, recognised him, and exhibited the most lively joy. On being set at liberty, the most affectionate Page 7 caresses were lavished on his old master, such as the most attached dog would have shown after an absence of a few days.

A second separation was followed by similar demonstrations of sorrow, which, however, again yielded to time. Three years passed, and the wolf was living happily in company with a dog, which had been placed with him, when his master again returned, and again the long-lost but still-remembered voice was instantly replied to by the most impatient cries, which were redoubled as soon as the poor animal was set at liberty; when, rushing to his master, he threw his fore-feet on his shoulders, licking his face with the most lively joy, and menacing his keepers, who offered to remove him, and towards whom, not a moment before, he had been showing every mark of fondness.

A third separation, however, seemed to be too much for this faithful animal's temper. He became gloomy, desponding, refused his food, and for a long time his life appeared in great danger. His health at last returned, but he no longer suffered the caresses of any but his keepers, and towards strangers manifested the original savageness of his species.

Mr. Bell, in his "History of Quadrupeds," mentions a curious fact, which, I think, still more strongly proves the alliance of the dog with the wolf, and is indeed exactly similar to what is frequently done by dogs when in a state of domestication. He informs Page 8 us, that he "remembers a bitch-wolf at the Zoological Gardens, which would always come to the front bars of her den to be caressed as soon as he, or any other person whom she knew, approached. When she had pups, she used to bring them in her mouth to be noticed; and so eager, in fact, was she that her little ones should share with her in the notice of her friends, that she killed all of them in succession by rubbing them against the bars of her den, as she brought them forwards to be fondled."

Other instances might be mentioned of the strong attachment felt by wolves to those who have treated them kindly, but I will now introduce some remarks on the anatomical affinities between the dog, the fox, and the wolf, which serve to prove that the dog is of a breed distinct from either of the last-mentioned animals.

It must, in fact, be always an interesting matter of inquiry respecting the descent of an animal so faithful to man, and so exclusively his associate and his friend, as the dog. Accordingly, this question has been entertained ever since Natural History took the rank of a science. But the origin of the dog is lost in antiquity. We find him occupying a place in the earliest pagan worship; his name has been given to one of the first-mentioned stars of the heavens, and his effigy may be seen in some of the most ancient works of art. Pliny was of opinion that there was no domestic animal without its unsubdued counterpart, and dogs are Page 9 known to exist absolutely wild in various parts of the old and new world. The Dingo of New Holland, a magnificent animal of this kind, has been shown to be susceptible of mutual attachment in a singular degree, though none of the experiments yet made have proved that he is capable, like the domestic dog, of a similar attachment to man. The parentage of the wild dogs has been assigned to the tame species, strayed from the dominion of their masters. This, however, still remains a question, and there is reason to believe that the wild dog is just as much a native of the wilderness as the lion or tiger. If there be these doubts about an animal left for centuries in a state of nature, how can we expect to unravel the difficulties accumulated by ages of domestication? Who knows for a certainty the true prototype of the goat, the sheep, or the ox? To the unscientific reader such questions might appear idle, as having been settled from time immemorial; yet they have never been finally disposed of. The difficulty, as with the dog, may be connected with modifications of form and colour, resulting from the long-continued interference of man with the breed and habits of animals subjected to his sway.

Buffon was very eloquent in behalf of the claim of the sheep-dog to be considered as the true ancestor of all the other varieties. Mr. Hunter would award this distinction to the wolf; supposing also that the jackal is the same animal a step further advanced towards Page 10 civilization, or perhaps the dog returned to its wild state.

As the affinity between wolf, jackal, fox, and dog, cannot fail to attract the notice of the most superficial observer; so he may ask if they do not all really belong to one species, modified by varieties of climate, food, and education? If answered in the negative, he would want to know what constitutes a species, little thinking that this question, apparently so simple, involves one of the nicest problems in natural history. Difference of form will scarcely avail us here, for the pug, greyhound, and spaniel, are wider apart in this respect, than many dogs and the wild animals just named. It has often been said that these varieties in the dog have arisen from artificial habits and breeding through a long succession of years. This seems very like mere conjecture. Can the greyhound be trained to the pointer's scent or the spaniel to the bulldog's ferocity? But admitting the causes assigned to be adequate to the effects, then the forms would be temporary, and those of a permanent kind only would serve our purpose. Of this nature is the shape of the pupil of the eye, which may be noticed somewhat particularly, not merely to make it plain to those who have never thought on the subject, but with the hope of leading them to reflections on this wondrous inlet to half our knowledge, the more especially as the part in question may be examined by any one in his own person by the help of a looking-glass. In the front of the eye then, just behind the transparent surface, there Page 11 is a sort of curtain called the *iris*, about the middle of which is a round hole. This is the pupil, and you will observe that it contracts in a strong light, and dilates in a weaker one, the object of which is to regulate the quantity of light admitted into the eye. Now the figure of the pupil is not the same in all animals. In the horse it is oval; in the wolf, jackal, and dog, it is round, like our own, however contracted; but in the fox, as in the cat, the pupil contracts vertically into an elongated figure, like the section of a lens, and even to a sort of slit, if the light be very strong.

This is a permanent character, not affected, as far as is at present known, by any artificial or natural circumstances to which the dog has been subjected. Naturalists, therefore, have seized upon this character as the ground for a division of animals of the dog kind, the great genus *Canis* of Linnæus, into two groups, the diurnal and nocturnal; not to imply that these habits necessarily belong to all the individuals composing either of these divisions, for that would be untrue, but simply that the figure of the pupils corresponds with that frequently distinguishing day-roaming animals from those that prowl only by night. It is remarkable that a more certain and serviceable specific distinction is thus afforded by a little anatomical point, than by any of the more obvious circumstances of form, size, or colour. Whether future researches into the minute structure of animals may not discover other means to assist the naturalist in distinguishing nearly Page 12 allied species, is a most important subject for inquiry, which cannot be entertained here. But to encourage those who may be disposed to undertake it, I must mention the curious fact, that the group to which the camel belongs is not more certainly indicated by his grotesque and singular figure than by the form of the red particles which circulate in his blood. And here again the inherent interest of the matter will lead me to enter a little into particulars, which may engage any one who has a good microscope in a most instructive course of observations, not the least recommendation of which is, that a just and pleasing source of recreation may be thus pursued by evening parties in the drawing-room, since the slightest prick of the finger will furnish blood enough for a microscopic entertainment, and you may readily procure a little more for comparison from any animal.

Now the redness of the blood is owing to myriads of minute objects in which the colour of the vital fluid resides. They were formerly called globules, but as they are now known to be flattened and disc-like, they are more properly termed particles or corpuscles. Their form is wonderfully regular, and so is their size within certain limits; in birds, reptiles, or fishes, the corpuscles are oval. They are circular in man, and all other mammalia, except in the camel tribe, in which the corpuscles are oval, though much smaller than in the lower animals. Thus, in the minutest drop of blood, any one of the camel family can be surely dis Page 13 tinguished from all other animals, even from its allies among the ruminants; and what is more to our purpose, in pursuing this inquiry, Mr. Gulliver has found that the blood-corpuscles of the dog and wolf agree exactly, while those of all the true foxes are slightly though distinctly smaller.

These curious facts are all fully detailed in Mr. Gulliver's Appendix to the English version of Gerber's Anatomy, but I think that they are now for the first time enlisted into the service of Natural History.

Thus we dismiss the fox as an alien to the dog, or, at all events, as a distinct species. Then comes the claim of the wolf as the true original of the dog. Before considering this, let us revert to the question of what constitutes a species. Mr. Hunter was of opinion that it is the power of breeding together and of continuing the breed with each other; that this is partially the case between the dog and the wolf is certain, for Lord Clanbrassil and Lord Pembroke proved the fact beyond a doubt, above half-a-century ago; and the following epitaph in the garden at Wilton House is a curious record of the particulars:—

Here lies Lupa,
Whose Grandmother was a Wolf,
Whose Father and Grandfather were Dogs, and whose
Mother was half Wolf and half Dog.
She died on the 16th of October, 1782,
Aged 12 years.

Conclusive as this fact may appear, as proving the Page 14 descent of the dog from the wolf, it is not convincing, the dog having characters which do not belong to the wolf.

The dog, for instance, guards property with strictest vigilance, which has been entrusted to his charge; all his energies seem roused at night, as though aware that that is the time when depredations are committed. His courage is unbounded, a property not possessed by the wolf: he appears never to forget a kindness, but soon loses the recollection of an injury, if received from the hand of one he loves, but resents it if offered by a stranger. His docility and mental pliability exceed those of any other animal; his habits are social, and his fidelity not to be shaken; hunger cannot weaken, nor old age impair it. His discrimination is equal, in many respects, to human intelligence. If he commits a fault, he is sensible of it, and shows pleasure when commended. These, and many other qualities, which might have been enumerated, are distinct from those possessed by the wolf. It may be said that domestication might produce them in the latter. This may be doubted, and is not likely to be proved; the fact is, the dog would appear to be a precious gift to man from a benevolent Creator, to become his friend, companion, protector, and the indefatigable agent of his wishes. While all other animals had the fear and dread of man implanted in them, the poor dog alone looked at his master with affection, and the tie once formed was never broken to the present hour.

Page 15 It should also be mentioned, in continuation of my argument, that the experiment of the wolf breeding with the dog is of no value, because it has never been carried sufficiently far to prove that the progeny would continue fertile *inter se*. The wolf has oblique eyes—the eyes of dogs have never retrograded to that position. If the dog descended from the wolf, a constant tendency would have been observed in the former to revert to the original type or species. This is a law in all other cross-breeds—but amongst all the varieties of dogs, this tendency has not existed. I may also add, that as far as I have been able to ascertain the fact, the number of teats of the female wolf have never been known to vary. With respect to the dog, it is known that they do vary, some having more, and others a less number.

Having thus brought forward such arguments as have occurred to me to prove that the dog is a breed *sui generis*, I will give a few anecdotes to show how different this animal is in his specific character to the wolf, and that he has a natural tendency to acknowledge man as his friend and protector, an instinct never shown by the wolf.

In Ceylon there are a great number of what are called wild dogs, that is, dogs who have no master, and who haunt villages and jungles, picking up what food they are able to find. If you meet one of these neglected animals, and only look at him with an expression of kindness, from that moment he attaches Page 16 himself to you, owns you for his master, and will remain faithful to you for the remainder of his life.

"Man," says Burns, "is the God of the dog; he knows no other; and see how he worships him! With what reverence he crouches at his feet, with what reverence he looks up to him, with what delight he fawns upon him, and with what cheerful alacrity he obeys him!"

Such is the animal which the brutality of man subjects to so much ill-treatment; its character depends very much on that of his master, kindness and confidence produce the same qualities in the dog, while ill-usage makes him sullen and distrustful of beings far more brutal than himself.

I have had many opportunities of observing how readily dogs comprehend language, and how they are aware when they are the subject of conversation. A gentleman once said in the hearing of an old and favourite dog, who was at the time basking in the sun,—"I must have Ponto killed, for he gets old and is offensive." The dog slunk away, and never came near his master afterwards. Many similar anecdotes might be brought forward, but I will mention one which Captain Brown tells us he received himself from Sir Walter Scott.

"The wisest dog I ever had," said Sir Walter, "was what is called the bulldog terrier. I taught him to understand a great many words, insomuch that Page 17 I am positive that the communication betwixt the canine species and ourselves might be greatly enlarged. Camp once bit the baker, who was bringing bread to the family. I beat him, and explained the enormity of his offence; after which, to the last moment of his life, he never heard the least allusion to the story, in whatever voice or tone it was mentioned, without getting up and retiring into the darkest corner of the room, with great appearance of distress. Then if you said, 'the baker was well paid,' or, 'the baker was not hurt after all,' Camp came forth from his hiding-place, capered, and barked, and rejoiced. When he was unable, towards the end of his life, to attend me when on horseback, he used to watch for my return, and the servant would tell him 'his master was coming down the hill, or through the moor,' and although he did not use any gesture to explain his meaning, Camp was never known to mistake him, but either went out at the front to go up the hill, or at the back to get down to the moor-side. He certainly had a singular knowledge of spoken language." An anecdote from Sir Walter Scott must be always pleasing.

Mr. Smellie, in his "Philosophy of Natural History," mentions a curious instance of the intellectual faculty of a dog. He states that "a grocer in Edinburgh had one which for some time amused and astonished the people in the neighbourhood. A man who went through the streets ringing a bell and selling pies, happened one day to treat this dog with a pie. The Page 18 next time he heard the pieman's bell he ran impetuously toward him, seized him by the coat, and would not suffer him to pass. The pieman, who understood what the animal wanted, showed him a penny, and pointed to his master, who stood at the street-door, and saw what was going on. The dog immediately supplicated his master by many humble gestures and looks, and on receiving a penny he instantly carried it in his mouth to the pieman, and received his pie. This traffic between the pieman and the grocer's dog continued to be daily practised for several months."

The affection which some dogs show to their masters and mistresses is not only very often surprising, but even affecting. An instance of this lately occurred at Brighton. The wife of a member of the town council at that place had been an invalid for some time, and at last was confined to her bed. During this period she was constantly attended by a faithful and affectionate dog, who either slept in her room or outside her door. She died, was buried, and the dog followed the remains of his beloved mistress to her grave. After the funeral the husband and his friends returned to the house, and while they were partaking of some refreshment the dog put its paws on his master's arm, as if to attract his attention, looked wistfully in his face, and then laid down and instantly expired.

In giving miscellaneous anecdotes in order to show the general character of the dog, I may mention the following very curious one.

Page 19 During a very severe frost and fall of snow in Scotland, the fowls did not make their appearance at the hour when they usually retired to roost, and no one knew what had become of them; the house-dog at last entered the kitchen, having in his mouth a hen, apparently dead. Forcing his way to the fire, the sagacious animal laid his charge down upon the warm hearth, and immediately set off. He soon came again with another, which he deposited in the same place, and so continued till the whole of the poor birds were rescued. Wandering about the stack-yard, the fowls had become quite benumbed by the extreme cold, and had crowded together, when the dog observing them, effected their deliverance, for they all revived by the warmth of the fire.

That dogs possess a faculty nearly allied to reason cannot, I think, be doubted. Mr. Davy, in his "Angler in the Lake District," (a charming work), gives one or two anecdotes in proof of this.

When Mr. Davy was at Ceylon, the Governor of that Island, the late Sir Robert Brownrigg, had a dog of more than ordinary sagacity. He always accompanied his master, being allowed to do so, except on particular occasions, such as going to church or council, or to inspect his troops, when the Governor usually wore his sword; but when the dog saw the sword girded on, he would only follow to the outer door. Without a word being said, he would return and wait the coming back of his master, patiently remaining Page 20 upstairs at the door of his private apartment. So it is with respect to my own pet terrier, Phiz. When he sees me putting on my walking-shoes, my great-coat, or hat, he is all eagerness to accompany me, jumping about me and showing his joy. But on Sundays it is very different. My shoes, great-coat or hat, may be put on, but he remains perfectly resigned on the rug before the fire, and never attempts or shows any inclination to follow me. Is the dog guided in acting thus by instinct or reason?

Let me give another instance from Mr. Davy's work.

Once when he was fishing in the highlands of Scotland, he saw a party of sportsmen, with their dogs, cross the stream, the men wading, the dogs swimming, with the exception of one, who stopped on the bank piteously howling. After a few minutes he suddenly ceased, and started off full speed for a higher part of the stream. Mr. Davy was able to keep him in view, and he did not stop till he came to a spot where a plank connected the banks, on which he crossed dry-footed, and soon joined his companions.

Dogs have sometimes strange fancies with respect to moving from one place to another. A Fellow of a College at Cambridge had a dog, which sometimes took it into his head to visit his master's usual places of resort in London. He would then return to his home in Suffolk, and then go to Cambridge, remaining at each place as long as he felt disposed to do so, and Page 21 going and returning with the most perfect indifference and complacency.

The extraordinary sense of a dog was shown in the following instance. A gentleman, residing near Pontypool, had his horse brought to his house by a servant. While the man went to the door, the horse ran away and made his escape to a neighbouring mountain. A dog belonging to the house saw this, and of his own accord followed the horse, got hold of the bridle and brought him back to the door.

I have been informed of two instances of dogs having slipped their collars and put their heads into them again of their own accord, after having committed depredations in the night, and I have elsewhere mentioned the fact of a dog, now in my possession, who undid the collar of another dog chained to a kennel near him. These are curious instances of sense and sagacity.

Mr. Bell, in his "History of British Quadrupeds," gives us the following fact of a dog belonging to a friend of his. This gentleman dropped a louis d'or one morning, when he was on the point of leaving his house. On returning late at night, he was told by his servant that the dog had fallen sick, and refused to eat, and, what appeared very strange, she would not suffer him to take her food away from before her, but had been lying with her nose close to the vessel, without attempting to touch it. On Mr. Bell's friend entering the room, the dog instantly jumped upon Page 22 him, laid the money at his feet, and began to devour her victuals with great voracity.

It is a curious fact that dogs can count time. I had, when a boy, a favourite terrier, which always went with me to church. My mother, thinking that he attracted too much of my attention, ordered the servant to fasten him up every Sunday morning. He did so once or twice, but never afterwards. Trim concealed himself every Sunday morning, and either met me as I entered the church, or I found him under my seat in the pew. Mr. Southey, in his "Omniana," informs us that he knew of a dog, which was brought up by a Catholic and afterwards sold to a Protestant, but still he refused to eat anything on a Friday.

Dogs have been known to die from excess of joy at seeing their masters after a long absence. An English officer had a large dog, which he left with his family in England, while he accompanied an expedition to America during the war of the Colonies. Throughout his absence, the animal appeared very much dejected. When the officer returned home, the dog, who happened to be lying at the door of an apartment into which his master was about to enter, immediately recognised him, leapt upon his neck, licked his face, and in a few minutes fell dead at his feet. A favourite spaniel of a lady recently died on seeing his beloved mistress after a long absence.

A gentleman who had a dog of a most endearing disposition, was obliged to go a journey periodically Page 23 once a-month. His stay was short, and his departure and return very regular, and without variation. The dog always grew uneasy when he first lost his master, and moped in a corner, but recovered himself gradually as the time for his return approached; which he knew to an hour, nay, to a minute. When he was convinced that his master was on the road, at no great distance from home, he flew all over the house; and if the street door happened to be shut, he would suffer no servant to have any rest until it was opened. The moment he obtained his freedom away he went, and to a certainty met his benefactor about two miles from town. He played and frolicked about him till he had obtained one of his gloves, with which he ran or rather flew home, entered the house, laid it down in the middle of the room, and danced round it. When he had sufficiently amused himself in this manner, out of the house he flew, returned to meet his master, and ran before him, or gambolled by his side, till he arrived with him at home. "I know not (says Mr. Dibdin, who relates this anecdote), how frequently this was repeated; but it lasted till the old gentleman grew infirm, and incapable of continuing his journeys. The dog by this time was also grown old, and became at length blind; but this misfortune did not hinder him from fondling his master, whom he knew from every other person, and for whom his affection and solicitude rather increased than diminished. The old gentleman, after a short illness, died. The dog knew the circumstance, Page 24 watched the corpse, blind as he was, and did his utmost to prevent the undertaker from screwing up the body in the coffin, and most outrageously opposed its being taken out of the house. Being past hope, he grew disconsolate, lost his flesh, and was evidently verging towards his end. One day he heard a gentleman come into the house, and he ran to meet him. His master being old and infirm, wore ribbed stockings for warmth. The gentleman had stockings on of the same kind. The dog perceived it, and thought it was his master, and began to exhibit the most extravagant signs of pleasure; but upon further examination finding his mistake, he retired into a corner, where in a short time he expired."

Some dogs are so faithful that they will never quit a thing entrusted to their charge, and will defend it to the utmost of their power. This may be often observed in the case of a cur, lying on the coat of a labourer while he is at work in the fields, and in those of carriers' and bakers' dogs. An instance is on record of a chimney-sweeper having placed his soot-bag in the street under the care of his dog, who suffered a cart to drive over and crush him to death, sooner than abandon his charge. Colonel Hamilton Smith, in the "Cyclopædia of Natural History," mentions a curious instance of fidelity and sagacity in a dog. He informs us that "in the neighbourhood of Cupar, in the county of Fife, there lived two dogs, mortal enemies to each other, and who always fought desperately whenever Page 25 they met. Capt. R— — was the master of one of them, and the other belonged to a neighbouring farmer. Capt. R— —'s dog was in the practice of going messages, and even of bringing butchers' meat and other articles from Cupar. One day, while returning charged with a basket containing some pieces of mutton, he was attacked by some of the curs of the town, who, no doubt, thought the prize worth contending for. The assault was fierce, and of some duration; but the messenger, after doing his utmost, was at last overpowered and compelled to yield up the basket, though not before he had secured a part of its contents. The piece saved from the wreck he ran off with, at full speed, to the quarters of his old enemy, at whose feet he laid it down, stretching himself beside it till he had eaten it up. A few snuffs, a few whispers in the ear, and other dog-like courtesies, were then exchanged; after which they both set off together for Cupar, where they worried almost every dog in the town; and, what is more remarkable, they never afterwards quarrelled, but were always on friendly terms."

That society and culture soften and moderate the passions of dogs cannot be doubted, and they constantly imbibe feelings from those of their master. Thus, if he is a coward, his dog is generally found to be one. Dogs are, however, in many respects, rational beings; and some proofs of this will be given in the present work. They will watch the countenance of Page 26 their master—they will understand words, which, though addressed to others, they will apply to themselves, and act accordingly. Thus a dog, which, from its mangy state, was ordered to be destroyed, took the first opportunity of quitting the ship, and would never afterwards come near a sailor belonging to it. If I desire the servant to wash a little terrier, who is apparently asleep at my feet, he will quit the room, and hide himself for some hours. A dog, though pressed with hunger, will never seize a piece of meat in presence of his master, though with his eyes, his movements, and his voice, he will make the most humble and expressive petition. Is not this reasoning?

But there is one faculty in the dog which would appear perfectly incomprehensible. It is the sense of smelling. He will not only scent various kinds of game at considerable distances, but he has been known to trace the odour of his master's feet through all the winding streets of a populous city. This extreme sensibility is very wonderful. It would thus appear that the feelings of dogs are more exquisite than our own. They have sensations, but their faculty of comparing them, or of forming ideas, is much circumscribed. A dog can imitate some human actions, and is capable of receiving a certain degree of instruction; but his progress soon stops. It is, however, an animal that should always be loved and treated with kindness. It is a curious fact, that dogs who have had their ears and tails Page 27 cut for many generations, transmit these defects to their descendants. Drovers' dogs, which may always be seen with short tails, are a proof of this.

A pleasing character of the dog is given in Smellie's "Philosophy of Natural History." He says:—

"The natural sagacity and talents of the dog are well known, and justly celebrated. But when these are improved by association with man, and by education, he becomes, in some measure, a rational being. The senses of the dog, particularly that of scenting distant objects, give him a superiority over every other quadruped. He reigns at the head of a flock; and his language, whether expressive of blandishment or of command, is better heard and better understood than the voice of his master. Safety, order, and discipline, are the effects of his vigilance and activity. Sheep and cattle are his subjects. These he conducts and protects with prudence and bravery, and never employs force against them except for the preservation of peace and good order. But when in pursuit of his prey, he makes a complete display of his courage and intelligence. In this situation both natural and acquired talents are exerted. As soon as the horn or voice of the hunter is heard, the dog demonstrates his joy by the most expressive emotions and accents. By his movements and cries he announces his impatience for combat, and his passion for victory. Sometimes he moves silently along, reconnoitres the ground, and endeavours to discover and surprise the enemy. At other times he Page 28 traces the animal's steps, and by different modulations of voice, and by the movements, particularly of his tail, indicates the distance, the species, and even the age of the fugitive deer. All these movements and modifications of voice are perfectly understood by experienced hunters. When he wishes to get into an apartment he comes to the door; if that is shut, he scratches with his foot, makes a bewailing noise, and, if his petition is not soon answered, he barks with a peculiar and humble voice. The shepherd's dog not only understands the language of his master, but, when too distant to be heard, he knows how to act by signals made with the hand."

Mr. Brockedon, in his "Journal of Excursions in the Alps," says:— "In these valleys, the early hours of retirement placed us in the difficult situation of fighting our way to the inn door at Lanslebourg against a magnificent Savoyard dog, who barked and howled defiance at our attempts, for which he stood some chance of being shot. At length a man, hearing our threats, popped his head out of a window, and entreated our forbearance. We were soon admitted, and refreshments amply provided. I had heard a story of a duel fought here from Mr. N— —, in which he was a principal, about a dog; and upon inquiry learnt that this was the same animal. A party of four young officers, returning from Genoa, stopped here. Mr. N— — had brought with him a beautiful little pet dog, which had been presented to him by a lady on his Page 29 leaving Genoa. Struck by the appearance of the fine dog at the inn, one of the officers bought it. He was fairly informed that the dog had been already sold to an Englishman, who had taken it as far as Lyons, where the dog escaped, and returned (two hundred miles) to Lanslebourg. The officer who made the purchase intended to fasten it in the same place with the little dog. This Mr. N— — objected to; when his brother-officer made some offensive allusions to the lady from whom the pet had been received. An apology was demanded, and refused. Swords were instantly drawn; they fought in the room. Mr. N— — wounded and disarmed his antagonist; an apology for the injurious reflections followed, and the party proceeded to England. The dog was taken safely as far as Paris, where he again escaped, and returned home (five hundred miles). I was now informed that the dog had been sold a third time to an Englishman; and again, in spite of precautions having been taken, he had returned to Lanslebourg from Calais."

A Scotch grazier, named Archer, having lost his way, and being benighted, at last got to a lone cottage; where, on his being admitted, a dog which had left Archer's house four years before immediately recognised him, fawned upon him, and when he retired for the night followed him into the chamber where he was to lie, and there, by his gestures, induced him narrowly to examine it; and then Archer saw sufficient to assure him that he was in the house of murderers. Rendered Page 30 desperate by the terrors of his situation, he burst into the room where the banditti were assembled, and wounded his insidious host by a pistol-shot; and in the confusion which the sudden explosion occasioned, he opened the door; and, notwithstanding he was fired at, accompanied by his dog Brutus, exerted all the speed which danger could call forth until daylight, which enabled him to perceive a house, and the main road, at no great distance. Upon his arrival at the house, and telling the master of it his story, he called up some soldiers that were there quartered, and who, by the aid of the dog, retraced the way back to the cottage. Upon examining the building a trap-door was found, which opened into a place where, amongst the mangled remains of several persons, was the body of the owner, who had received the shot from the grazier's pistol in his neck; and although not dead, had been, by the wretches his associates, in their quick retreat, thrown into this secret cemetery. He was, however, cured of his wound, delivered up to justice, tried, and executed. [A]

A merchant had received a large sum of money; and being fatigued with riding in the heat of the day, had retired to repose himself in the shade; and upon remounting his horse, had forgotten to take up the bag which contained the money. His dog tried to remind his master of his inadvertency by crying and barking, which so surprised the merchant, that, in crossing a Page 31 brook, he observed whether the dog drank, as he had his suspicions of his being mad; and which were confirmed by the dog's not lapping any water, and by his increased barking and howling, and at length by his endeavouring to bite the heels of the horse. Impressed with the idea of the dog's madness, to prevent further mischief, he discharged his pistol at him, and the dog fell. After riding some distance with feelings that will arise in every generous breast at the destruction of an affectionate animal, he discovered that his money was missing. His mind was immediately struck that the actions of the dog, which his impetuosity had construed into madness, were only efforts to remind him of his loss. He galloped back to where he had fired his pistol; but the dog was gone from thence with equal expedition to the spot where he had reposed. But what were the merchant's feelings when he perceived his faithful dog, in the struggles of death, lying by the side of the bag which had been forgotten! The dog tried to rise, but his strength was exhausted. He stretched out his tongue to lick the hand that was now fondling him with all the agony of regret for the wound its rashness had inflicted, and casting a look of kindness on his master, closed his eyes for ever. [B]

I am indebted to a well-known sportsman for the following interesting account of some of his dogs. It affords another proof how much kindness will do in bringing out the instinctive faculties of these animals; Page 32 and that, when properly educated, their sense, courage, and attachment are most extraordinary.

"Smoaker was a deer greyhound of the largest size, but of his pedigree I know nothing. In speed he was equal to any hare greyhound; at the same time, in spirit he was indomitable. He was the only dog I ever knew who was a match for a red stag, single-handed. From living constantly in the drawing-room, and never being separated from me, he became acquainted with almost the meaning of every word—certainly of every sign. His retrieving of game was equal to any of the retrieving I ever saw in any other dogs. He would leap over any of the most dangerous spikes at a sign, walk up and come down any ladder, and catch, without hurting it, any particular fowl out of a number that was pointed out to him. If he missed me from the drawing-room, and had doubts about my being in the house, he would go into the hall and look for my hat: if he found it, he would return contented; but if he did not find it, he would proceed up-stairs to a window at the very top of the house, and look from the window each way, to ascertain if I were in sight. One day in shooting at Cranford, with his late Royal Highness the Duke of York, a pheasant fell on the other side of the stream. The river was frozen over; but in crossing to fetch the pheasant the ice broke, and let Smoaker in, to some inconvenience. He picked up the pheasant, and instead of trying the ice again, he took it many hundred yards round to the Page 33 bridge. Smoaker died at the great age of eighteen years. His son Shark was also a beautiful dog. He was by Smoaker out of a common greyhound bitch, called Vagrant, who had won a cup at Swaffham. Shark was not so powerful as Smoaker; but he was, nevertheless, a large-sized dog, and was a first-rate deer greyhound and retriever. He took his father's place on the rug, and was inseparable from me. He was educated and entered at deer under Smoaker. When Shark was first admitted to the house, it chanced that one day he and Smoaker were left alone in a room with a table on which luncheon was laid. Smoaker might have been left for hours with meat on the table, and he would have died rather than have touched it; but at that time Shark was not proof against temptation. I left the room to hand some lady to her carriage, and as I returned by the window, I looked in. Shark was on his legs, smelling curiously round the table; whilst Smoaker had risen to a sitting posture, his ears pricked, his brow frowning, and his eyes intently fixed on his son's actions. After tasting several viands, Shark's long nose came in contact with about half a cold tongue; the morsel was too tempting to be withstood. For all the look of curious anger with which his father was intently watching, the son stole the tongue and conveyed it to the floor. No sooner had he done so, than the offended sire rushed upon him, rolled him over, beat him, and took away the tongue. Instead, though, of replacing it on the table, Page 34 the father contented himself with the punishment he had administered, and retired with great gravity to the fire.

"I was once waiting by moonlight for wild ducks on the Ouze in Bedfordshire, and I killed a couple on the water at a shot. The current was strong; but Shark, having fetched one of the birds, was well aware there was another. Instead, therefore, of returning by water to look for the second, he ran along the banks, as if aware that the strong stream would have carried the bird further down; looking in the water till he saw it, at least a hundred yards from the spot where he had left it in bringing the first; when he also brought that to me. Nothing could induce either of these dogs to fetch a glove or a stick: I have often seen game fall close to me, and they would not attempt to touch it. It seemed as if they simply desired to be of service when service was to be done; and that when there were no obstacles to be conquered, they had no wish to interfere. Shark died at a good old age, and was succeeded by his son Wolfe. Wolfe's mother was a Newfoundland bitch. He was also a large and powerful dog, but of course not so speedy as his ancestors. While residing at my country house, being my constant companion, Wolfe accompanied me two or three times a-day in the breeding season to feed the young pheasants and partridges reared under hens. On going near the coops, I put down my gun, made Wolfe a sign to sit down by it, and fed the birds, with some Page 35 caution, that they might not be in any way scared. I mention this, because I am sure that dogs learn more from the manner and method of those they love, than they do from direct teaching. In front of the windows on the lawn there was a large bed of shrubs and flowers, into which the rabbits used to cross, and where I had often sent Wolfe in to drive them for me to shoot. One afternoon, thinking that there might be a rabbit, I made Wolfe the usual sign to go and drive the shrubs, which he obeyed; but ere he had gone some yards beneath the bushes, I heard him make a peculiar noise with his jaws, which he always made when he saw anything he did not like, and he came softly back to me with a sheepish look. I repeated the sign, and encouraged him to go; but he never got beyond the spot he had been to in the first instance, and invariably returned to me with a very odd expression of countenance. Curiosity tempted me to creep into the bushes to discover the cause of the dog's unwonted behaviour; when there, I found, congregated under one of the shrubs, eight or nine of my young pheasants, who had for the first time roosted at a distance from their coop. Wolfe had seen and known the young pheasants, and would not scare them.

"Wolfe was the cause of my detecting and discharging one of my gamekeepers. I had forbidden my rabbits to be killed until my return; and the keeper was ordered simply to walk Wolfe to exercise on the farm. There was a large stone quarry in the Page 36 vicinity, where there were a good many rabbits, some parts of which were so steep, that though you might look over the cliff, and shoot a rabbit below, neither man nor dog could pick him up without going a considerable way round. On approaching the edge of the quarry to look over for a rabbit, I was surprised at missing Wolfe, who invariably stole off in another direction, but always the same way. At last, on shooting a rabbit, I discovered that he invariably went to the only spot by which he could descend to pick up whatever fell to the gun; and by this I found that somebody had shot rabbits in his presence at times when I was from home.

"Wolfe accompanied me to my residence in Hampshire, and there I naturalised, in a wild state, some white rabbits. For the first year the white ones were never permitted to be killed, and Wolfe saw that such was the case. One summer's afternoon I shot a white rabbit for the first time, and Wolfe jumped the garden fence to pick the rabbit up; but his astonishment and odd sheepish look, when he found it was a white one, were curious in the extreme. He dropped his stern, made his usual snap with his jaws, and came back looking up in my face, as much as to say, 'You've made a mistake, and shot a white rabbit, but I've not picked him up.' I was obliged to assure him that I intended to shoot it, and to encourage him before he would return and bring the rabbit to me. Wolfe died when he was about nine years old, and was succeeded Page 37 by my present favourite, Brenda, a hare greyhound of the highest caste. Brenda won the Oak stakes of her year, and is a very fast and stout greyhound. I have taught her to retrieve game to the gun, to drive home the game from dangerous sands, and, in short, to do everything but speak; and this she attempts, by making a beautiful sort of bark when she wants her dinner.

"I have the lop-eared rabbit naturalised, and in a half-wild and wild state, and Brenda is often to be seen with some of the tamest of them asleep in the sun on the lawn together. When the rabbits have been going out into a dangerous vicinity, late in the evening, I have often sent Brenda to drive them home, and to course and kill the wild ones if she could. I have seen one of the wild-bred lop-ears get up before her, and I have seen her make a start to course it; but when she saw that it was not a native of the soil she would stop and continue her search for others. The next moment I have seen her course and kill a wild rabbit. She is perfectly steady from hare if I tell her not to run, and is, without any exception, one of the prettiest and most useful and engaging creatures ever seen. She is an excellent rat-killer also, and has an amazing antipathy to a cat. When I have been absent from home for some time, Mrs. B. has observed that she is alive to every sound of a wheel, and if the door-bell rings she is the first to fly to it. When walking on the sea-beach during my absence, she is Page 38 greatly interested in every boat she sees, and watches them with the most intense anxiety, as in the yachting season she has known me return by sea. Brenda would take my part in a row, and she is a capital house-dog. If ever the heart of a creature was given to man, this beautiful, graceful, and clever animal has given me hers, for her whole existence is either passed in watching for my return, or in seeking opportunities to please me when I am at home. It is a great mistake to suppose that severity of treatment is necessary to the education of a dog, or that it is serviceable in making him steady. Manner—*marked and impressive manner*—is that which teaches obedience, and example rather than command forms the desired character.

"I had two foxhounds when I hunted stag,—my pack were all foxhounds,—they were named Bachelor and Blunder. We used to play with them together, and they got to know each other by name. In returning from hunting, my brother and myself used to amuse ourselves by saying, in a peculiar tone of voice,—the one we used to use in playing with them—'Bachelor, where's Blunder?' On hearing this, Bachelor's stern and bristles rose, and he trotted about among the pack, looking for Blunder, and when he found him he would push his nose against his ear and growl at him. Thus Bachelor evidently knew Blunder by name, and this arose from the way in which we used to play with them. At this moment, when far away from home, and after an absence of Page 39 many weeks, if I sing a particular song, which I always sing to a dog named Jessie, Brenda, though staying in houses where she had never seen Jessie, will get up much excited, and look to the door and out of the window in expectation of her friend. I have a great pleasure in the society of all animals, and I love to make my house a place where all may meet in rest and good fellowship. This is far easier to achieve than people would think for when dogs are kindly used, but impressed with ideas of obedience.

"The gazelle which came home from Acre in the Thunderer, was one evening feeding from Mrs. B.'s plate at dessert, when Odion, the great deerhound, who was beaten in my match against the five deer by an unlucky stab in the first course, came in by special invitation for his biscuit. The last deer he had seen previous to the gazelle he had coursed and pulled down. The strange expression of his dark face was beautiful when he first saw her; and halting in his run up to me, he advanced more slowly directly to her, she met him also in apparent wonder at his great size, and they smelled each others' faces. Odion then kissed her, and came to me for his biscuit, and never after noticed her. She will at times butt him if he takes up too much of the fire; but this she will not do to Brenda, except in play; and if she is eating from Mrs. Berkeley's hand, Brenda by a peculiar look can send her away and take her place. Odion, the gazelle, Brenda, and the rabbits, will all quietly lay on the Page 40 lawn together, and the gazelle and Bruiser, an immense house-dog between the bloodhound and mastiff, will run and play together.

"I had forgotten to mention a bull-and-mastiff dog that I had, called Grumbo. He was previous to Smoaker, and was indeed the first four-footed companion established in my confidence. I was then very young, and of course inclined to anything like a row. Grumbo, therefore, was well entered in all kinds of strife—bulls, oxen, pigs, men, dogs, all came in turn as combatants; and Grumbo had the oddest ways of making men and animals the *aggressors* I ever knew. He seemed to make it a point of honour never to begin, but on receiving a hint from me; some one of his enemies was sure to commence the battle, and then he or both of us would turn to as an oppressed party. I have seen him walk leisurely out into the middle of a field where oxen were grazing, and then throw himself down. Either a bull or the oxen were sure to be attracted by the novel sight, and come dancing and blowing round him. All this he used to bear with the most stoical fortitude, till some one more forward than the rest touched him with the horn. 'War to the knife, and no favour,' was then the cry; and Grumbo had one of them by the nose directly. He being engaged at odds, I of course made in to help him, and such a scene of confusion used to follow as was scarce ever seen. Grumbo tossed in the air, and then some beast pinned by the nose would lie down Page 41 and bellow. I should all this time be swinging round on to some of their tails, and so it would go on till Grumbo and myself were tired and our enemies happy to beat a retreat. If he wished to pick a quarrel with a man, he would walk listlessly before him till the man trod on him, and then the row began. Grumbo was the best assistant, night or day, for catching delinquents, in the world. As a proof of his thoughtful sagacity, I give the following fact. He was my sole companion when I watched two men steal a quantity of pheasants' eggs: we gave chase; but before I could come near them, with two hundred yards start of me, they fled. There was no hope of my overtaking them before they reached the village of Harlington, so I gave Grumbo the office. Off he went, but in the chase the men ran up a headland on which a cow was tethered. They passed the cow; and when the dog came up to the cow he stopped, and, to my horror, contemplated a grab at the tempting nose. He was, however, uncertain as to whether or not this would be right, and he looked back to me for further assurance. I made the sign to go ahead, and he understood it, for he took up the running again, and disappeared down a narrow pathway leading through the orchards to the houses. When I turned that corner, to my infinite delight I found him placed in the narrow path, directly in front of one of the poachers, with such an evident determination of purpose, that the man was standing stock still, afraid to stir either hand or foot. I came Page 42 up and secured the offender, and bade the dog be quiet."

It is, I believe, a fact, and if so, it is a curious one, that the dog in a wild state only howls; but when he becomes the friend and companion of man, he has then wants and wishes, hopes and fears, joys and sorrows, to which in his wilder state he appears to have been a stranger. His vocabulary, if it may be so called, then increases, in order to express his enlarged and varying emotions. He anticipates rewards and punishments, and learns to solicit the former and deprecate the latter. He bounds exultingly forth to accompany his master in his walks, rides, and sports of the field. He acts as the faithful guardian of his property. He is his fire-side companion, evidently discerns days of household mirth or grief, and deports himself accordingly. Hence, his energies and his sensibilities are all expanded, and what he feels he seeks to tell in various accents, and in different ways. For instance, our little dog comes and pulls his mistress's gown and makes significant whines, if any one is in or about the premises whom he thinks has no right to be there. I have seen a dog pick up a stick and bring it in his mouth to his master, looking at the water first and then at his master, evidently that the stick might be thrown into it, that he might have the pleasure of swimming after it. In my younger days, I was in the habit of teazing a favourite dog by twitching his nose and pretending to pull his ears. He would snap gently at me, but if, Page 43 by accident, he gave me rather a harder bite than he had intended, he became instantly aware of it, and expressed his regret in a way not to be mistaken. Dogs who have hurt or cut themselves will submit patiently while the wound is being dressed, however much the operation may hurt them. They become instantly sensible that no punishment is intended to be inflicted, and I have seen them lick the hand of the operator, as if grateful for what he was doing. Those who are in the habit of having dogs constantly in the room with them, will have perceived how alive they are to the slightest change in the countenance of their master; how gently they will touch him with their paw when he is eating, in order to remind him of their own want of food; and how readily they distinguish the movements of any inmate of the house from those of a stranger. These, and many other circumstances which might be mentioned, show a marked distinction between a domesticated dog and one that is wild, or who has lived with people who are in an uncivilized state, such as the Esquimaux, &c. Both the wild and domestic dog, however, appear to be possessed of and to exercise forethought. They will bury or hide food, which they are unable to consume at once, and return for it. But the domestic dog, perhaps, gives stronger proofs of forethought; and I will give an instance of it. A large metal pot, turned on one side, in which a great quantity of porridge had been boiled, was set before a Newfoundland puppy of three or four months old. At first, he Page 44 contented himself by licking off portions of the oatmeal which adhered to the interior, but finding this unsatisfactory, he scraped the morsels with his fore-paws into a heap, and then ate the whole at once. I had a dog, who, having once scalded his tongue, always afterwards, when I gave him his milk and water at breakfast, put his paw very cautiously into the saucer, to see if the liquid was too hot, before he would touch it with his tongue.

Dogs have frequently been known to hunt in couples; that is, to assist each other in securing their prey: thus associating together and admitting of no partnership.

At Palermo, in Sicily, there is an extraordinary quantity of dogs wandering about without owners. Amongst the number, two more particularly distinguished themselves for their animosity to cats. One day they were in pursuit of a cat, which, seeing no other place of refuge near, made her escape into a long earthen water-pipe which was lying on the ground. These two inseparable companions, who always supported each other, pursued the cat to the pipe, where they were seen to stop, and apparently to consult each other as to what was to be done to deceive and get possession of the poor cat. After they had stood a short time they divided, taking post at each end of the pipe, and began to back alternately, thus giving the cat reason to suppose that they were both at one end, in order to induce her to come out. This manœuvre had a successful result, and the cheated cat left her hiding-place. Page 45 Scarcely had she ventured out, when she was seized by one of the dogs; the other hastened to his assistance, and in a few moments deprived her of life. [C]

The memory of dogs is quite extraordinary, and only equalled by that of the elephant. Mr. Swainson, in his work on the instincts of animals, gives the following proof of this. He says that "A spaniel belonging to the Rev. H. N., being always told that he must not follow his master to church on Sundays, used on those days to set off long before the service, and lie concealed under the hedge, so near the church, that at length the point was yielded to him." My little parlour dog never offers to go with me on a Sunday, although on other days he is perfectly wild to accompany me in my walks.

In my younger days I had a favourite dog, which always accompanied me to church. My mother, seeing that he attracted too much of my attention, ordered the servant to shut him every Sunday morning. This was done once, but never afterwards; for he concealed himself early every Sunday morning, and I was sure to find him either under my seat at church, or else at the church-door. That dogs clearly distinguish the return of Sunday cannot be doubted.

The almost incredible penetration and expedition with which dogs are known to return to their former homes, from places to which they have been sent, or Page 46 carried in such a recluse way as not to retain a trace of the road, will ever continue to excite the greatest admiration.

A dog having been given by a gentleman at Wivenhoe to the captain of a collier, he took the dog on board his vessel, and landed him at Sunderland; but soon after his arrival there the dog was missing, and in a very few days arrived at the residence of his old master, in Essex. A still more extraordinary circumstance is upon record, of the late Colonel Hardy, who, having been sent for express to Bath, was accompanied by a favourite spaniel bitch in his chaise, which he never quitted till his arrival there. After remaining there four days, he accidentally left his spaniel behind him, and returned to his residence at Springfield, in Essex, with equal expedition; where, in three days after, his faithful and steady adherent arrived also, notwithstanding the distance between that place and Bath is 140 miles, and she had to explore her way through London, to which she had never been, except in her passage to Bath, and then within the confines of a close carriage. [D]

In the small town of Melbourne, in Derbyshire, cocks and hens may be seen running about the streets. One day a game cock attacked a small bantam, and they fought furiously, the bantam having, of course, the worst of it. Some persons were standing about looking at the fight, when my informant's house-dog suddenly darted out, snatched up the bantam in his mouth, and carried it Page 47 into the house. Several of the spectators followed, believing that the poor fowl would be killed and eaten by the dog; but his intentions were of a more benevolent nature. After guarding the entrance of the kennel for some time, he trotted down the yard into the street, looked about to the right and left, and seeing that the coast was clear, he went back again, and once more returning with his *protégé* in his mouth, safely deposited him in the street, and then walked quietly away. How few human beings would have acted as this dog had done!

Here is another curious anecdote from Mr. Davy's work. He says that the cook in the house of a friend of his, a lady on whose accuracy he could rely, and from whom he had the anecdote, missed a marrow-bone. Suspicion fell on a well-behaved dog—a great favourite, and up to that time distinguished for his honesty. He was charged with the theft; he hung down his tail, and for a day or two was altered in his manner, having become shy, sullen, and sheepish, to use these expressions for want of better. In this mood he continued, till, to the amusement of the cook, he brought back the bone and laid it at her feet. Then, with the restoration of her stolen property, he resumed his cheerful manner. How can we interpret this conduct of the dog, better than by supposing that he was aware he had done amiss, and that the evil doing preyed on him till he had made restitution? Was not this a kind of moral sense?

If a dog finds a bone while he is accompanying his master in a walk, he does not stay behind to gnaw it, Page 48 but runs some distance in advance, attacks the bone, waits till his master comes up, and then proceeds forward again with it. By acting in this manner, he never loses sight of his master.

A dog has been known to convey food to another of his species who was tied up and pining for want of it. A dog has frequently been seen to plunge voluntarily into a rapid stream, to rescue another that was in danger of drowning. He has defended helpless curs from the attacks of other dogs, and learns to apportion punishment according to the provocation received, frequently disdaining to exercise his power and strength on a weaker adversary. Repeated provocation will, however, excite and revenge. For instance, a Newfoundland dog was quietly eating his mess of broth and broken scraps. While so employed, a turkey endeavoured to share the meal with him. The dog growled, and displayed his teeth. The intruder retired for a moment, but quickly returned to the charge, and was again "warned off," with a like result. After three or four attempts of the same kind, the dog became provoked, gave a sudden ferocious growl, bit off the delinquent's head, and then quietly finished his meal, without bestowing any further attention on his victim.

The celebrated Leibnitz related to the French Academy an account of a dog he had seen which was taught to speak, and could call in an intelligible manner for tea, coffee, chocolate, Page 49 &c.

The dog was of a middling size, and the property of a peasant in Saxony. A little boy, the peasant's son, imagined that he perceived in the dog's voice an indistinct resemblance to certain words, and was, therefore, determined to teach him to speak distinctly. For this purpose he spared neither time nor pains with his pupil, who was about three years old when his learned education commenced; and at length he made such progress in language, as to be able to articulate no less than thirty words. It appears, however, that he was somewhat of a truant, and did not very willingly exert his talents, being rather pressed into the service of literature, and it was necessary that the words should be first pronounced to him each time before he spoke. The French Academicians who mention this anecdote, add, that unless they had received the testimony of so great a man as Leibnitz, they should scarcely have dared to relate the circumstance.

An invalid gentleman, who resided for some years on Ham Common, in Surrey, had a dog which distinctly pronounced John, William, and two or three other words. A medical friend of mine, who attended this gentleman, has frequently heard the animal utter these words; and a female relative of his, who was often on a visit at his house, assures me of the fact. Indeed it need not be doubted.

These are the only two instances I have met with of talking dogs, but my brother had a beautiful little spaniel, named Doll, who was an indefatigable hunter Page 50 after woodcocks and snipes. Doll would come home in the evening after a hard day's sport, wet, tired and dirty, and then deposit herself on the rug before the fire. Happening one day to pull her ear gently when in this state, she expressed her dislike to be disturbed by a sort of singing noise. By repeating this from day to day, and saying "Sing, Doll," she would utter notes of a somewhat musical tone, and continue for some time after I had ceased to touch her ear, to the amusement and surprise of those who heard her. Poor Doll! I shall never see your like again, either for beauty or intelligence. If she was affronted she would come to me, at a distance of four miles, remain some time, and then return to her master.

A small cur, blind of one eye, lame, ugly, old, and somewhat selfish, yet possessed of great shrewdness, was usually fed with three large dogs. Watching his opportunity, he generally contrived to seize the best bit of offal or bone, with which he retreated into a recess, the opening to which was so small that he knew the other dogs could not follow him into it, and where he enjoyed his repast without the fear of molestation.

Early habits predominate strongly in dogs, and indeed in other animals. At the house of a gentleman in Wexford, out of four dogs kept to guard the premises, three of them would always wag their tails, and express what might be called civility, on the approach Page 51 of any well-dressed visitors; manifesting, on the other hand, no very friendly feelings towards vagrants or ill-dressed people. The fourth,—a sort of fox-hound,—which, as a puppy, had belonged to a poor man, always seemed to recognise beggars and ill-dressed passengers as old familiar friends, growling at well-attired strangers, barking vehemently at gigs, and becoming almost frantic with rage at a four-wheeled carriage.

The olfactory nerves of a dog are quite extraordinary, and it is said that, making allowance for difference of corporeal bulk, they are about four times larger than those of a man. Some dogs, however, seem to excel in acuteness of hearing, and others in peculiar powers of vision.

We quote the following from the "Percy Anecdotes:"—

"One day, when Dumont, a tradesman of the Rue St. Denis, was walking in the Boulevard St. Antoine with a friend, he offered to lay a wager with the latter, that if he were to hide a six-livre piece in the dust, his dog would discover and bring it to him. The wager was accepted, and the piece of money secreted, after being carefully marked. When the two had proceeded some distance from the spot, M. Dumont called to his dog that he had lost something, and ordered him to seek it. Caniche immediately turned back, and his master and his companion pursued their walk to the Rue St. Denis. Meanwhile a traveller, who happened Page 52 to be just then returning in a small chaise from Vincennes, perceived the piece of money, which his horse had kicked from its hiding-place; he alighted, took it up, and drove to his inn, in the Rue Pont-aux-Choux. Caniche had just reached the spot in search of the lost piece when the stranger picked it up. He followed the chaise, went into the inn, and stuck close to the traveller. Having scented out the coin which he had been ordered to bring back in the pocket of the latter, he leaped up incessantly at and about him. The traveller, supposing him to be some dog that had been lost or left behind by his master, regarded his different movements as marks of fondness; and as the animal was handsome, he determined to keep him. He gave him a good supper, and on retiring to bed took him with him to his chamber. No sooner had he pulled off his breeches, than they were seized by the dog; the owner conceiving that he wanted to play with them, took them away again. The animal began to bark at the door, which the traveller opened, under the idea that the dog wanted to go out. Caniche snatched up the breeches, and away he flew. The traveller posted after him with his night-cap on, and literally *sans culottes*. Anxiety for the fate of a purse full of gold Napoleons, of forty francs each, which was in one of the pockets, gave redoubled velocity to his steps. Caniche ran full speed to his master's house, where the stranger arrived a moment afterwards, breathless and enraged. He accused the dog of robbing him. 'Sir,' said the Page 53 master, 'my dog is a very faithful creature; and if he has run away with your breeches, it is because you have in them money which does not belong to you.' The traveller became still more exasperated. 'Compose yourself, sir,' rejoined the other, smiling; 'without doubt there is in your purse a six-livre piece, with such and such marks, which you have picked up in the Boulevard St. Antoine, and which I threw down there with the firm conviction that my dog would bring it back again. This is the cause of the robbery which he has committed upon you.' The stranger's rage now yielded to astonishment; he delivered the six-livre piece to the owner, and could not forbear caressing the dog which had given him so much uneasiness, and such an unpleasant chase."

A gentleman in Cornwall possessed a dog, which seemed to set a value on white and shining pebble stones, of which he had made a large collection in a hole under an old tree. A dog in Regent Street is said to have barked with joy on hearing the wheels of his master's carriage driven to the door, when he could not by any possibility see the vehicle, and while many other carriages were at the time passing and repassing. This, I believe, is a fact by no means uncommon.

My retriever will carry an egg in his mouth to a great distance, and during a considerable length of time, without ever breaking or even cracking the shell. A small bird having escaped from its cage and fallen Page 54 into the sea, a dog conveyed it in his mouth to the ship, without doing it the slightest injury.

RETRIEVER.

One of the carriers of a New York paper called the "Advocate," having become indisposed, his son took his place; but not knowing the subscribers he was to supply, he took for his guide a dog which had usually attended his father. The animal trotted on a-head of the boy, and stopped at every door where the paper was in use to be left, without making a single omission or mistake.

The following is from a newspaper of this year: Page 55 –

"A most extraordinary circumstance has just occurred at the Hawick toll-bar, which is kept by two old women. It appears that they had a sum of money in the house, and were extremely alarmed lest they should be robbed of it. Their fears prevailed to such an extent, that, when a carrier whom they knew was passing by, they urgently requested him to remain with them all night, which, however, his duties would not permit him to do; but, in consideration of the alarm of the women, he consented to leave with them a large mastiff dog. In the night the women were disturbed by the uneasiness of the dog, and heard a noise apparently like an attempt to force an entrance into the premises, upon which they escaped by the back-door, and ran to a neighbouring house, which happened to be a blacksmith's shop. They knocked at the door, and were answered from within by the smith's wife. She said her husband was absent, but that she was willing to accompany the terrified women to their home. On reaching the house, they heard a savage but half-stifled growling from the dog. On entering they saw the body of a man hanging half in and half out of their little window, whom the dog had seized by the throat, and was still worrying. On examination, the man proved to be their neighbour the blacksmith, dreadfully torn about the throat, and quite dead."

A dog, belonging to the late Dr. Robert Hooper, had been in the constant habit of performing various little personal services for his master, such as fetching Page 56 his slippers, &c. It happened one day that Dr. Hooper had been detained by his professional duties much beyond his usual dinner hour. The dog impatiently waited for his arrival, and he at last returned, weary and hungry. After showing his pleasure at the arrival of his master, greeting him with his usual attention, the animal remained tolerably quiet until he conceived a reasonable time had elapsed for the preparation of the Doctor's dinner. As it did not, however, make its appearance, the dog went into the kitchen, seized with his mouth a half-broiled beefsteak, with which he hastened back to his master, placing it on the table-cloth before him.

A few years ago, the public were amused with an account given in the newspapers of a dog which possessed the strange fancy of attending all the fires that occurred in the metropolis. The discovery of this predilection was made by a gentleman residing a few miles from town, who was called up in the middle of the night by the intelligence that the premises adjoining his house of business were on fire. "The removal of my books and papers," said he, in telling the story, "of course claimed my attention; yet, notwithstanding this, and the bustle which prevailed, my eye every now and then rested on a dog, which, during the hottest progress of the conflagration, I could not help noticing running about, and apparently taking a deep interest in what was going on; contriving to keep himself out of everybody's way, and yet always present amidst the Page 57 thickest of the stir. When the fire was got under, and I had leisure to look about me, I again observed the dog, which, with the firemen, appeared to be resting from the fatigues of duty, and was led to make some inquiries respecting him. 'Is this your dog, my friend?' said I to a fireman. 'No, sir,' answered he; it does not belong to me, or to any one in particular. We call him the firemen's dog.' 'The firemen's dog!' I replied. 'Why so? Has he no master?' 'No, sir,' rejoined the fireman; 'he calls none of us master, though we are all of us willing enough to give him a night's lodging and a pennyworth of meat. But he won't stay long with any of us. His delight is to be at all the fires in London; and, far or near, we generally find him on the road as we are going along, and sometimes, if it is out of town, we give him a lift. I don't think there has been a fire for these two or three years past which he has not been at.'

"The communication was so extraordinary, that I found it difficult to believe the story, until it was confirmed by the concurrent testimony of several other firemen. None of them, however, were able to give any account of the early habits of the dog, or to offer any explanation of the circumstances which led to this singular propensity.

"Some time afterwards, I was again called up in the night to a fire in the village in which I resided (Camberwell, in Surrey), and to my surprise here I again met with 'the firemen's dog,' still alive and Page 58 well, pursuing, with the same apparent interest and satisfaction, the exhibition of that which seldom fails to bring with it disaster and misfortune, oftentimes loss of life and ruin. Still, he called no man master, disdained to receive bed or board from the same hand more than a night or two at a time, nor could the firemen trace out his resting-place."

Such was the account of this interesting animal as it appeared in the newspapers, to which were shortly afterwards appended several circumstances communicated by a fireman at one of the police offices. A magistrate having asked him whether it was a fact that the dog was present at most of the fires that occurred in the metropolis, the fireman replied that he never knew "Tyke," as he was called, to be absent from a fire upon any occasion that he (the fireman) attended himself. The magistrate said the dog must have an extraordinary predilection for fires. He then asked what length of time he had been known to possess that propensity. The fireman replied that he knew Tyke for the last nine years; and although he was getting old, yet the moment the engines were about, Tyke was to be seen as active as ever, running off in the direction of the fire. The magistrate inquired whether the dog lived with any particular fireman. The fireman replied that Tyke liked one fireman as well as another; he had no particular favourites, but passed his time amongst them, sometimes going to the house of one, and then to another, and off to a third when he was tired. Day Page 59 or night, it was all the same to him; if a fire broke out, there he was in the midst of the bustle, running from one engine to another, anxiously looking after the firemen; and, although pressed upon by crowds, yet, from his dexterity, he always escaped accidents, only now and then getting a ducking from the engines, which he rather liked than otherwise. The magistrate said that Tyke was a most extraordinary animal; and having expressed a wish to see him, he was shortly after exhibited at the office, and some other peculiarities respecting him were related. There was nothing at all particular in the appearance of the dog; he was a rough-looking small animal, of the terrier breed, and seemed to be in excellent condition, no doubt from the care taken of him by the firemen belonging to the different companies. There was some difficulty experienced in bringing him to the office, as he did not much relish going any distance from where the firemen are usually to be found, except in cases of attending with them at a conflagration, and then distance was of no consequence. It was found necessary to use stratagem for the purpose. A fireman commenced running. Tyke, accustomed to follow upon such occasions, set out after him; but this person, having slackened his pace on the way, the sagacious animal, knowing there was no fire, turned back, and it was necessary to carry him to the office.

The following striking anecdote, of a similar kind, appeared in the first number of the new issue of Page 60 Cassell's "Illustrated Family Paper." After giving a short account of a fire-escape man, named Samuel Wood, the writer thus alludes to his dog Bill:—

"As to Bill, he regards him evidently in the light of a friend; he had him when he was a pup from a poor fellow who died in the service, and he and his 'Bill' have been on excellent terms ever since.

"The fire-escape man's dog takes after his master in courage and perseverance. He is of the terrier breed, six years old. An alarm of fire calls forth all his energy. He is the first to know that something is wrong—the first to exert himself in setting it right. He has not been trained to the work—'it is a gift,' as his master says; and if we all used our gifts as efficiently as the dog Bill, it would be the better for us. On an alarm of fire Bill barks his loudest, dashes about in a frantic manner, till his master and the escape are on their way to it. He, of course, is there first, giving the police and the crowd to understand that Wood and his fire-escape are coming. When the escape is fixed, and Wood begins to ascend the ladder, Bill runs up the canvas; as soon as a window is opened, Bill leaps in and dashes about to find the occupants, loudly barking for assistance as soon as he has accomplished his errand of mercy. His watchfulness and sagacity are never at fault, although on more than one occasion he has stood a fair chance of losing his life, and has sustained very severe injury. Not long ago a collar was presented to Bill as a reward for his Page 61 services; unfortunately for him, he has since lost this token of public regard—a misfortune much to be regretted. The following verse was engraved on the collar:—

> 'I am the fire-escape man's dog: my name is Bill.
> When 'fire' is called I am never still:
> I bark for my master, all danger brave,
> To bring the escape—human life to save.'

Collared or collarless, Bill is always ready to lend a helping bark. May his life be long, and his services properly esteemed!"

The following anecdote shows extraordinary sense, if not reasoning faculty, in a dog:—

A lady of high rank has a sort of colley, or Scotch sheep-dog. When he is ordered to ring the bell, he does so; but if he is told to ring the bell when the servant is in the room whose duty it is to attend, he refuses, and then the following occurrence takes place. His mistress says, "Ring the bell, dog." The dog looks at the servant, and then barks his bow wow, once or twice. The order is repeated two or three times. At last the dog lays hold of the servant's coat in a significant manner, just as if he had said to him—"Don't you hear that I am to ring the bell for you?—come to my lady." His mistress always had her shoes warmed before she put them on, but one day during the hot weather her maid was putting them on without their having been previously placed before the fire. When the dog saw this he immediately interfered, expressing the greatest indignation at the maid's negli Page 62 gence. He took the shoes from her, carried them to the fire, and after they had been warmed as usual, he brought them back to his mistress with much apparent satisfaction, evidently intending to say, if he could, "It is all right now."

The dispositions and characters of dogs, as well as their intelligence, vary very much. Let me give a few instances of this.

When that benevolent man, Mr. Backhouse, went to Australia, in hopes of doing good among the convicts, he was residing in the house of a gentleman who had a son about four years of age. This boy strayed one morning into the bush, and could not be found after a long search had been made for him. In the evening a little dog, which had accompanied the child, scratched at the door, and on its being opened showed unmistakeable signs of wishing to be followed. This was done; and he led the way to the child, who was at last found sitting by the side of a river three or four miles from the house.

At Albany in Worcestershire, at the seat of Admiral Maling, a dog went every day to meet the mail, and brought the bag in his mouth to the house. The distance was about a half-a-quarter of a mile. The dog usually received a meal of meat as his reward. The servants having, on *one day only*, neglected to give him his accustomed meal, the dog on the arrival of the next mail buried the bag, nor was it found without considerable search.

Page 63 M. D'Obsonville had a dog which he had brought up in India from two months old; and having to go with a friend from Pondicherry to Bengalore, a distance of more than nine hundred miles, he took the animal along with him. "Our journey," says M. D'O., "occupied nearly three weeks; and we had to traverse plains and mountains, and to ford rivers, and go along by-paths. The animal, which had certainly never been in that country before, lost us at Bengalore, and immediately returned to Pondicherry. He went directly to the house of my friend, M. Beglier, then commandant of artillery, and with whom I had generally lived. Now the difficulty is not so much to know how the dog subsisted on the road (for he was very strong, and able to procure himself food), but how he should so well have found his way after an interval of more than a month! This was an effort of memory greatly superior to that which the human race is capable of exerting."

A gentleman residing in Denmark, Mr. Decouick, one of the king's privy councillors, found that he had a remarkable dog. It was the habit of Mr. Decouick to leave Copenhagen on Fridays for Drovengourd, his country seat. If he did not arrive there on the Friday evening, the dog would invariably be found at Copenhagen on Saturday morning, in search of his master. Hydrophobia becoming common, all dogs were shot that were found running about, an exception being made in the case of Mr. Decouick's dog on account of Page 64 his sagacity and fidelity, a distinctive mark being placed upon him.

The following anecdotes are from Daniel's "Rural Sports:" —

Upon the fidelity of dogs, the following facts deserve to be here recorded: of this property, or other peculiar traits, if they appertain to any class of sporting dogs, in that class they will be noticed.

Dr. Beattie, in one of his ingenious and elegant essays, relates a story, in his own knowledge, of a gentleman's life being saved, who fell beneath the ice, by his dog's going in quest of assistance, and almost forcibly dragging a farmer to the spot.

Mr. Vaillant describes the losing of a bitch while travelling in Africa, when after firing his gun, and fruitlessly searching for her, he despatched one of his attendants, to return by the way they had proceeded; when she was found at about two leagues' distance, seated by the side of a chair and basket, which had dropped unperceived from his waggon: an instance of attentive fidelity, which must have proved fatal to the animal, either from hunger or beasts of prey, had she not been luckily discovered.

As instances of the dog's sagacity, the following are submitted. In crossing the mountain St. Gothard, near Airola, the Chevalier Gaspard de Brandenberg and his servant were buried by an avalanche; his dog, who escaped the heap of snow, did not quit the place where he had lost his master: this was, fortunately, not far Page 65 from the convent; the animal howled, ran to the convent frequently, and then returned. Struck by his perseverance, the next morning the people from the house followed him; he led them directly to the spot, scratched the snow, and after thirty-six hours passed beneath it, the chevalier and his domestic were taken out safe, hearing distinctly during their confinement the howling of the dog and the discourse of their deliverers. Sensible that to the sagacity and fondness of this creature he owed his life, the gentleman ordered by his will that he should be represented on his tomb with his dog; and at Zug, in the church of St. Oswald, where he was buried in 1728, they still show the monument and the effigy of this gentleman, with the dog lying at his feet.

In 1792, a gentleman, who lived in Vere Street, Clare Market, went with his family to the pit of Drury Lane Theatre, at about half-past five in the evening, leaving a small spaniel, of King Charles's breed, locked up in the dining-room, to prevent the dog from being lost in his absence. At eight o'clock his son opened the door, and the dog immediately went to the playhouse and found out his master, though the pit was unusually thronged, and his master seated near its centre.

A large dog of Mr. Hilson's, of Maxwelhaugh, on the 21st of October, 1797, seeing a small one that was following a cart from Kelso carried by the current of the Tweed, in spite of all its efforts to bear up against Page 66 the stream, after watching its motions attentively, plunged voluntarily into the river, and seizing the tired animal by the neck, brought it safely to land.

The docility of the dog is such, that he may be taught to practise with considerable dexterity a variety of human actions: to open a door fastened by a latch, and pull a bell when desirous to be admitted. Faber mentions one belonging to a nobleman of the Medici family, which always attended at its master's table, took from him his plates, and brought him others; carried wine to him in a glass upon a salver, which it held in its mouth, without spilling; the same dog would also hold the stirrup in its teeth while its master was mounting his horse. Mr. Daniel had formerly a spaniel, which he gave the honourable Mr. Greville, that, beyond the common tricks which dogs trained to fetch and carry exhibit, would bring the bottles of wine from the corner of the room to the table by the neck, with such care as never to break one; and, in fact, was the *boots* of the mess-room.

Some few years since, the person who lived at the turnpike-house, about a mile from Stratford-upon-Avon, had trained a dog to go to the town for any small parcels of grocery, &c. which he wanted. A note, mentioning the things required, was tied round his neck, and in the same manner the articles were fastened, and arrived safe to his master.

Colonel Hutchinson relates the following anecdote: Page 67 –

"A cousin of one of my brother-officers was taking a walk at Tunbridge Wells, when a strange Newfoundland snatched her parasol from her hand, and carried it off. The lady followed the dog, who kept ahead, constantly looking back to see if she followed. The dog at length stopped at a confectioner's, and went in, followed by the lady, who, as the dog would not resign it, applied to the shopman for assistance. He then told her that it was an old trick of the dog's to get a bun, and that if she would give him one he would return the property. She cheerfully did so, and the dog as willingly made the exchange."

The above anecdote proves that dogs are no mean observers of countenances, and that he had satisfied himself by a previous scrutiny as to the probability of his delinquencies being forgiven.

Of the abstinence and escape of a dog, the following narrative may not be uninteresting: —

In 1789, when preparations were making at St. Paul's for the reception of his majesty, a favourite dog followed its master up the dark stairs of the dome. Here, all at once, it was missing; and calling and whistling were to no purpose. Nine weeks after this, all but two days, some glaziers were at work in the cathedral, and heard a faint noise amongst the timbers which support the dome. Thinking it might be some unfortunate human being, they tied a rope round a boy, and let him down near the place whence the sound came. At the bottom he found a dog lying on its Page 68 side, the skeleton of another dog, and an old shoe half eaten. The humanity of the boy led him to rescue the animal from its miserable situation, and it was accordingly drawn up. Much emaciated, and scarce able to stand, the workmen placed it in the porch of the church, to die or live as it might happen. This was about ten o'clock in the morning. Some time after, the dog was seen endeavouring to cross the street at the top of Ludgate Hill; but its weakness was so great, that, unsupported by a wall, it could not accomplish it. The miserable appearance of the dog again excited the compassion of a boy, who carried it over. By the aid of the houses it was enabled to get to Fleet Market, and over two or three narrow crossings in its way to Holborn Bridge, and about eight o'clock in the evening it reached its master's house in Red Lion Street, Holborn, and laid itself down on the steps, having been ten hours in its journey from St. Paul's to that place. The dog was so much altered, its eyes being so sunk in its head as to be scarce discernible, that the master would not encourage his faithful old companion, who when lost was supposed to weigh twenty pounds, but now only weighed three pounds fourteen ounces. The first indication it gave of knowing its master was by wagging its tail when he mentioned its name, Phillis; for a long time it was unable to eat or drink, and it was kept alive by the sustenance it received from its mistress, who used to feed it with a teaspoon. At length it recovered. It must not be supposed that this Page 69 animal existed for nine weeks without food; she was in whelp when lost, and doubtless ate her young. The remains of another dog, killed by a similar fall, were likewise found, and were most probably converted by the survivor to the most urgent of all natural purposes; and when this treat was done, the shoe succeeded, which was almost half devoured. What famine and a thousand accidents could not do, was effected a short time after by the wheels of a coach, which unfortunately went over her, and ended the life of poor Phillis.

Of dogs that have supported themselves in a wild state, to the great loss and annoyance of the farmer, there are two instances worthy of notice, from the cunning with which both these dogs frustrated, for a length of time, every secret and open attack. In December, 1784, a dog was left by a smuggling vessel near Boomer, on the coast of Northumberland. Finding himself deserted, he began to worry sheep, and did so much damage that he was the terror of the country, within the circuit of above twenty miles. It is asserted, that when he caught a sheep, he bit a hole in its right side, and after eating the fat about the kidneys, left it. Several of them, thus lacerated, were found alive by the shepherds; and being properly taken care of, some of them recovered, and afterwards had lambs. From this delicacy of his feeding, the destruction may in some measure be conceived, as the fat of one sheep in a day would scarcely satisfy his Page 70 hunger. Various were the means used to destroy him: frequently was he pursued with hounds, greyhounds, &c., but when the dogs came up with him, he laid down on his back, as if supplicating for mercy, and in that position they never hurt him; he therefore laid quietly, taking his rest, until the hunters approached, when he made off without being followed by the hounds, until they were again excited to the pursuit, which always terminated unsuccessfully. He was one day pursued from Howick to upwards of thirty miles' distance, but returned thither and killed sheep the same evening. His constant residence was upon a rock on the Heugh Hill, near Howick, where he had a view of four roads that approached it; and there, in March 1785, after many fruitless attempts, he was at last shot.

Another wild dog, which had committed similar devastation among the sheep, near Wooler, in the same county (Northumberland), was, on the 6th of June, 1799, advertised to be hunted on the Wednesday following, by three packs of hounds, which were to meet at different places; the aid of men and fire-arms was also requested, with a reward promised of twenty guineas to the person killing him. This dog was described by those who had seen him at a distance as a large greyhound, with some white in his face, neck and one fore-leg white, rather grey on the back, and the rest of a jet-black. An immense concourse of people assembled at the time appointed, but Page 71 the chase was unprosperous; for he eluded his pursuers among the Cheviot Hills, and, what is singular, returned that same night to the place from whence he had been hunted in the morning, and worried an ewe and her lamb. During the whole summer he continued to destroy the sheep, but changed his quarters, for he infested the fells, sixteen miles south of Carlisle, where upwards of sixty sheep fell victims to his ferocity. In September, hounds and firearms were again employed against him, and after a run from Carrock Fell, which was computed to be thirty miles, he was shot whilst the hounds were in pursuit by Mr. Sewel of Wedlock, who laid in ambush at Moss Dale. During the chase, which occupied six hours, he frequently turned upon the headmost hounds, and wounded several so badly as to disable them. Upon examination, he appeared of the Newfoundland breed, of a common size, wire-haired, and extremely lean. This description does not tally with the dog so injurious to the farmers in Northumberland, although, from circumstances, there is little doubt but it was the same animal.

With a laughably philosophical account of dogs, under the supposition of a transmigration of souls, and with their general natural history from Linnæus and Buffon, this introductory chapter will be concluded.

A facetious believer in the art of distinguishing at the sight of any creature from what class of animals his soul is derived, thus allots them:—

The souls of deceased bailiffs and common con Page 72 stables are in the bodies of setting dogs and pointers; the terriers are inhabited by trading justices; the bloodhounds were formerly a set of informers, thief-takers, and false evidences; the spaniels were heretofore courtiers, hangers-on of administrations, and hack journal-writers, all of whom preserve their primitive qualities of fawning on their feeders, licking their hands, and snarling and snapping at all who offer to offend their master; a former train of gamblers and black-legs are now embodied in that species of dog called lurchers;

bull-dogs and mastiffs were once butchers and drovers; greyhounds and hounds owe their animation to country squires and foxhunters; little whiffling, useless lap-dogs, draw their existence from the quondam beau; macaronies, and gentlemen of the tippy, still being the playthings of ladies, and used for their diversion. There are also a set of sad dogs derived from attornies; and puppies, who were in past time attornies' clerks, shopmen to retail haberdashers, men-milliners, &c. &c. Turnspits are animated by old aldermen, who still enjoy the smell of the roast meat; that droning, snarling species, styled Dutch pugs, have been fellows of colleges; and that faithful, useful tribe of shepherds' dogs, were, in days of yore, members of parliament, who guarded the flock, and protected the sheep from wolves and thieves, although indeed of late some have turned sheep-biters, and worried those they ought to have defended.

Linnæus informs us, the dog eats flesh, and Page 73 farinaceous vegetables, but not greens, (this is a mistake, for they will eat greens when boiled); its stomach digests bones; it uses the tops of grass as a vomit; is fond of rolling in carrion; voids its excrements on a stone; its dung (the *album græcum*) is one of the greatest encouragers of putrefaction; it laps up its drink with its tongue; makes water sideways, by lifting up one of its hind-legs; is most diuretic in the company of a strange dog, and very apt to repeat it where another dog has done the same: *Odorat anum alterius, menstruans catulit cum variis; mordet illa illos; cohæret copula junctus*. Its scent is most exquisite when its nose is moist; it treads lightly on its toes; scarce ever sweats, but when hot, lolls out its tongue; generally walks frequently round the place it intends to lie down on; its sense of hearing is very quick when asleep; it dreams. It goes with young sixty-three days, and commonly brings from four to ten; the male puppies resemble the dog, the female the bitch (an assertion by no means accurate, any more than the tail always bending to the left is a common character of the species). It is the most faithful of animals, is very docile, fawns at his master's approach, runs before him on a journey, often passing over the same ground; on coming to crossways, stops and looks back; drives cattle home from the field; keeps herds and flocks within bounds, protects them from wild beasts; points out to the sportsman the game; brings the birds that are shot to its master; will turn a spit; at Brussels, Page 74 and in Holland, draws little carts to the herb-market; in more northern regions, draws sledges with provisions, travellers, &c.; will find out what is dropped; watchful by night, and when the charge of a house or garden is at such times committed to him, his boldness increases, and he sometimes becomes perfectly ferocious; when it has been guilty of a theft, slinks away with its tail between its legs; eats voraciously, with oblique eyes; enemy to beggars; attacks strangers without provocation; hates strange dogs; howls at certain notes in music, and often urines on hearing them; will snap at a stone thrown at it; is sick at the approach of bad weather, (a remark vague and uncertain); is afflicted with worms; spreads its madness; grows blind with age; *sæpe gonorrhæâ infectus*; driven as unclean from the houses of the Mahometans; yet the same people establish hospitals for, and allow them daily food.

The dog, says Buffon, like every other animal which produces above one or two at a time, is not perfectly formed immediately after birth. Dogs are always brought forth blind; the two eyelids are not simply glued together, but shut up with a membrane, which is torn off, as soon as the muscles of the upper eyelids acquire strength sufficient to overcome this obstacle to vision, which generally happens the tenth or twelfth day. At this period, the bones of the head are not completed, the body and muzzle are bloated, and the whole figure is ill defined; but in less than two months, they learn to use all their senses; their Page 75 growth is rapid, and they soon gain strength. In the fourth month, they lose some of their teeth, which, as in other animals, are soon replaced, and never again fall out: they have six cutting and two canine teeth in each jaw, and fourteen grinders in the upper, and twelve in the under, making in all forty-two teeth; but the number of grinders sometimes varies in particular dogs.

The time of gestation is nine weeks, or sixty-three days; sometimes sixty-two or sixty-one, but never less than sixty.

The bitch produces six, seven, and even so far as twelve puppies, and generally has more at the subsequent litters than she has at the first; but the observation of Buffon, that a female hound, covered by a dog of her own kind, and carefully shut up from all others, has been known to produce a mixed race, consisting of hounds and terriers, is totally void of foundation. A curious circumstance, in the account of the setter, will be mentioned, of an impression made upon the mind of a bitch of that sort by the attention of a cur, which never had access to her, and yet her whelps were always like him, and possibly this hound bitch had a violent hankering after some terrier.

Dogs continue to propagate during life, which is commonly limited to fourteen or fifteen years, yet some have been known to exceed twenty, but that is rare. The duration of life in this, as in other animals, bears propor Page 76 tion to the time of his growth, which in the dog is not completed in less than two years, and he generally lives fourteen. His age may be discovered by his teeth; when young, they are white, sharp, and pointed; as he increases in years, they become black, blunt, and unequal: it may likewise be known by the hair, which turns grey on the muzzle, front, and round the eyes.

The manner in which the shepherds of the Pyrenees employ their peculiar breed of dogs, which are large, long-haired, of a tawny white colour, and a very strong build, with a ferocious temper, exhibits a vivid instance of the trust they repose in the courage and fidelity of these animals, and of the virtues by which they merit and reward it. Attended by three or more dogs, the shepherds will take their numerous flocks at early dawn to the part of the mountain side which is destined for their pasture. Having counted them, they descend to follow other occupations, and commit the guardianship of the sheep to the sole watchfulness of the dogs. It has been frequently known, that when wolves have approached, the three sentinels would walk round and round the flock, gradually compressing them into so small a circle that one dog might with ease overlook and protect them, and that this measure of caution being executed, the remaining two would set forth to engage the enemy, over whom, it is said, they invariably triumph.

The following interesting remarks are extracted from Chambers: Page 77 –

The educability of the dog's perceptive faculties has been exemplified in a remarkable manner by his acquired knowledge of musical sounds. On some dogs fine music produces an apparently painful effect, causing them gradually to become restless, to moan piteously, and, finally, to fly from the spot with every sign of suffering and distress. Others have been seen to sit and listen to music with seeming delight, and even to go every Sunday to church, with the obvious purpose of enjoying the solemn and powerful strains of the organ. Some dogs manifest a keen sense of false notes in music. Mrs. Samuel Carter Hall, at Old Brompton, possesses an Italian greyhound, which screams in apparent agony when a jarring combination of notes is produced, accidentally or intentionally, on the piano. These opposite and various manifestations show what might be done by education to teach dogs a critical knowledge of sounds. A gentleman of Darmstadt, in Germany, as we learn, has taught a poodle dog to detect false notes in music. We give the account of this remarkable instance of educability as it appears in a French newspaper.

Mr. S— —, having acquired a competency by commercial industry, retired from business, and devoted himself, heart and soul, to the cultivation and enjoyment of music. Every member of his little household was by degrees involved more or less in the same occupation, and even the housemaid could in time bear a part in a chorus, or decipher a melody of Schubert. Page 78 One individual alone in the family seemed to resist this musical entrancement; this was a small spaniel, the sole specimen of the canine race in the mansion. Mr. S— — felt the impossibility of instilling the theory of sounds into the head of Poodle, but he firmly resolved to make the animal bear *some* part or other in the general domestic concert; and by perseverance, and the adoption of ingenious means, he attained his object. Every time that a *false note* escaped either from the instrument or voice—as often as any blunder, of whatever kind, was committed by the members of the musical family (and such blunders were sometimes committed intentionally)—down came its master's cane on the back of the unfortunate poodle, till she howled and growled again. Poodle perceived the meaning of these unkind chastisements, and instead of becoming sulky, showed every disposition to howl on the instant a false note was uttered, without waiting for the formality of a blow. By and by, a mere glance of Mr. S— —'s eye was sufficient to make the animal howl to admiration. In the end, Poodle became so thoroughly acquainted with, and attentive to, false notes and other musical barbarisms, that the slightest mistake of the kind was infallibly signalised by a yell from her, forming the most expressive commentary upon the misperformance.

When extended trials were made of the animal's acquirements, they were never found to fail, and Poodle became, what she still is, the most famous, impartial, and conscientious connoisseur in the Duchy of Hesse. Page 79 But, as may be imagined, her musical appreciation is entirely negative; if you sing with expression, and play with ability, she will remain cold and impassible. But let your execution exhibit the slightest defect, and you will have her instantly showing her teeth, whisking her tail, yelping, barking, and growling. At the present time, there is not a concert or an opera at Darmstadt to which Mr. S— — and his wonderful dog are not invited; or, at least, *the dog*. The voice of the prima donna, the instruments of the band—whether violin, clarionet, hautbois, or bugle—all of them must execute their parts in perfect harmony, otherwise Poodle looks at its master, erects its ears, shows its grinders, and howls outright. Old or new pieces, known or unknown to the dog, produce on it the same effect.

It must not be supposed that the discrimination of the creature is confined to the mere *execution* of musical compositions. Whatever may have been the case at the outset of its training, its present and perfected intelligence extends even to the secrets of composition. Thus, if a vicious modulation, or a false relation of parts, occur in a piece of music, the animal shows symptoms of uneasy hesitation; and if the error be continued, will infallibly give the grand condemnatory howl. In short, Poodle is the terror of all the middling composers of Darmstadt, and a perfect nightmare to the imagination of all poor singers and players. Sometimes Mr. S— — and his friends take a pleasure in annoying the canine critic, by emitting all sorts of Page 80 discordant sounds from instrument and voice. On such occasions the creature loses all self-command, its eyes shoot forth fiery flashes, and long and frightful howls respond to the immelodious concert of the mischievous bipeds. But the latter must be careful not to go too far; for when the dog's patience is tried to excess, it becomes altogether wild, and flies fiercely at the tormentors and their instruments.

This dog's case is a very curious one, and the attendant phenomena not very easy of explanation. From the animal's power of discerning the correctness of musical composition, as well as of execution, one would be inclined to imagine that Mr. S— —, in training his dog, had only called into play faculties existing (but latent) before, and that dogs have in them the natural germs of a fine musical ear. This seems more likely to be the case, than that the animal's perfect musical taste was wholly an acquirement, resulting from the training. However this may be, the Darmstadt dog is certainly a marvellous creature, and we are surprised that, in these exhibiting times, its powers have not been displayed on a wider stage. The operatic establishments of London and Paris might be greatly the better, perhaps, for a visit from the critical Poodle.

It is now settled, as a philosophical question, that the instruction communicated to dogs, as well as various other animals, has an hereditary effect on the progeny. If a dog be taught to perform certain feats, Page 81 the young of that dog will be much easier initiated in the same feats than other dogs. Thus, the existing races of English pointers are greatly more accomplished in their required duties than the original race of Spanish pointers. Dogs of the St. Bernard variety inherit the faculty of tracking footsteps in the snow. A gentleman of our acquaintance, and of scientific acquirements, obtained some years ago a pup, which had been produced in London by a female of the celebrated St. Bernard breed. The young animal was brought to Scotland, where it was never observed to give any particular tokens of a power of tracking footsteps until winter, when the ground became covered with snow. It *then* showed the most active inclination to follow footsteps; and so great was its power of doing so under these circumstances, that, when its master had crossed a field in the most curvilinear way, and caused other persons to cross his path in all directions, it nevertheless followed his course with the greatest precision. Here was a perfect revival of the habit of its Alpine fathers, with a degree of specialty as to external conditions at which, it seems to us, we cannot sufficiently wonder.

Such are some of the qualities of dogs in a state of domestication, and let me hope that the anecdotes related of them will tend to insure for them that love and gratitude to which their own fine disposition and noble character give them a claim from us.

It is pleasing to observe that men of the highest Page 82 acquirements and most elevated minds have bestowed their sincere attachment upon their favourite canine companions; for kindness to animals is, perhaps, as strong an indication of the possession of generous sentiments as any that can be adduced. The late Lord Grenville, a distinguished statesman, an elegant scholar, and an amiable man, affords an illustration of the opinion: It is thus that he eloquently makes his favourite Zephyr speak:—

> "Captum oculis, senioque hebetem, morboque gravatum,
> Dulcis here, antiquo me quod amore foves,
> Suave habet et carum Zephyrus tuus, et leviore
> Se sentit mortis conditione premi.
> Interiêre quidem, tibi quæ placuisse solebant,
> Et formæ dotes, et facile ingenium:
> Deficiunt sensus, tremulæ scintillula vitæ
> Vix micat, in cinerem mox abitura brevem.
> Sola manet, vetuli tibi nec despecta ministri,
> Mens grata, ipsaque in morte memor domini.
> Hanc tu igitur, pro blanditiis mollique lepore,
> Et prompta ad nutus sedulitate tuos,
> Pro saltu cursuque levi, lusuque protervo,
> Hanc nostri extremum pignus amoris habe.
> Jamque vale! Elysii subeo loca læta, piorum
> Quæ dat Persephone manibus esse canum."

In the previous pages I have endeavoured to give my readers some idea of the general character of the dog, and I will now proceed to illustrate it more fully by anecdotes peculiar to different breeds. These animals will then be found to deserve the encomiums bestowed upon them by Buffon, "as possessing such an ardour of sentiment, with fidelity and constancy in their affec Page 83 tion, that neither ambition, interest, nor desire of revenge, can corrupt them, and that they have no fear but that of displeasing. They are, in fact, all zeal, ardour, and obedience. More inclined to remember benefits than injuries; more docile and tractable than any other animal, the dog is not only instructed, but conforms himself to the manners, movements, and habits of those who govern him. He is always eager to obey his master, and will defend his property at the risk of his own life." Pope says, that history is more full of examples of fidelity in the dog than in friends; and Lord Byron characterises him as—

> "in life the firmest friend,
> The first to welcome, foremost to defend;
> Whose honest heart is still his master's own;
> Who labours, fights, lives, breathes for him alone;"

and truly indeed may he be called

> "The rich man's guardian, and the poor man's friend."

Page 84

Page 85

"His bulk and beauty speak no vulgar praise.
*　　*　　*　　*
Oh had you seen him, vigorous, bold, and young,
Swift as a stag, and as a lion strong;
Him no fell savage in the plain withstood,
None 'scap'd him, bosomed in the gloomy wood;
His eye how piercing!"—Pope.

THE IRISH AND HIGHLAND WOLF-DOG.

A certain degree of romance will always be attached to the history of the Irish wolf-dog, but so contradictory are the accounts handed down to us respecting it, Page 86 that, with every disposition to do justice to the character of this noble animal, the task is one of no small difficulty.

This dog seems to have flourished, and to have become nearly extinct, with the ancient kings of Ireland, and, with the harp and shamrock, is regarded as one of the national emblems of that country. When princely hospitality was to be found in the old palaces, castles, and baronial halls of fair Erin, it is hardly possible to imagine anything more aristocratic and imposing than the aspect of these dogs, while attending the banquets of their masters. So great, indeed, was their height, that it has been affirmed, that when their chieftain was seated at table these dogs could rest their heads on his shoulders. However this may have been, it is certain that the bold, majestic, and commanding appearance of the animal, joined to the mild and softened look with which he regarded those to whom he was attached, and whom he was always ready to defend, must have rendered him worthy of the enthusiasm with which the remembrance of him is still cherished by the warm-hearted people of Ireland.

The following anecdote, which has been communicated to me by an amiable Irish nobleman, will at all events serve to show the peculiar instinct which the Irish wolf-dog was supposed to possess.

A gentleman of an ancient family, whose name it is unnecessary to mention, from his having been engaged in the troubles which agitated Ireland about fifty Page 87 or sixty years since, went into a coffee-room at Dublin during that period, accompanied by a noble wolf-dog, supposed to be one of the last of the breed. There was only one other gentleman in the coffee-room, who, on seeing the dog, went up to him, and began to notice him. His owner, in considerable alarm, begged him to desist, as the dog was fierce, and would never allow a stranger to touch him. The gentleman resumed his seat, when the dog came to him, showed the greatest pleasure at being noticed, and allowed himself to be fondled. His owner could not disguise his astonishment. "You are the only person," he said, "whom that dog would ever allow to touch him without showing resentment. May I beg of you the favour to tell me your name?"— mentioning his own at the same time. The stranger announced it, (he was the last of his race, one of the most ancient and noble in Ireland, and descended from one of its kings.) "I do not wonder," said the owner of the dog, "at the homage this animal has paid to you. He recognizes in you the descendant of one of our most ancient race of gentlemen to whom this breed of dogs almost exclusively belonged, and the peculiar instinct he possesses has now been shown in a manner which cannot be mistaken by me, who am so well acquainted with the ferocity this dog has hitherto shown to all strangers."

Few persons, Sir Walter Scott excepted, would perhaps be inclined to give credit to this anecdote. So convinced was he of the extraordinary instinct Page 88 exhibited by dogs generally, that he has been heard to declare that he would believe anything of a dog. The anecdote, however, above related, was communicated to me with the strongest assurance of its strict accuracy.

In a poem, written by Mrs. Catherine Philips, about the year 1660, the character of the Irish wolf-hound is well portrayed, and proves the estimation in which he was held at that period.

"Behold this creature's form and state!
Him Nature surely did create,
That to the world might be exprest
What mien there can be in a beast;
More nobleness of form and mind
Than in the lion we can find:
Yea, this heroic beast doth seem
In majesty to rival him.

Yet he vouchsafes to man to show
His service, and submission too—
And here we a distinction have;
That brute is fierce—the dog is brave.

He hath himself so well subdued,
That hunger cannot make him rude;

And all his manners do confess
That courage dwells with gentleness.

War with the wolf he loves to wage,
And never quits if he engage;
But praise him much, and you may chance
To put him out of countenance.
And having done a deed so brave,
He looks not sullen, yet looks grave.

Page 89

No fondling play-fellow is he;
His master's guard he wills to be:
Willing for him his blood be spent,
His look is never insolent.
Few men to do such noble deeds have learn'd,
Nor having done, could look so unconcern'd."

This is one of the finest descriptions of a noble dog which I have yet met with in English poetry. Courage and modesty are well portrayed, and contrasted.

The following anecdotes relate to an animal which must have strongly resembled the Irish wolf-dog:—

Plutarch mentions a certain Roman in the civil wars, whose head nobody durst cut off for fear of the dog that guarded his body, and fought in his defence. The same author relates that King Pyrrhus, in the course of one of his journies, observed a dog watching over a dead body; and hearing that he had been there three days without meat or drink, ordered the body to be buried, and the dog taken care of and brought to him. A few days afterwards there was a muster of the soldiers, so that every man had to march in order before the king. The dog lay quiet for some time; but when he saw the murderers of his late master pass by, he flew upon them with extraordinary fury, barking, and tearing their garments, and frequently turning about to the king; which both excited the king's suspicion, and that of all who stood about him. The men were in consequence apprehended, and though the circumstances which appeared in evidence against Page 90 them were very slight, they confessed the crime, and were accordingly punished.

Montfaucon mentions a similar case of attachment and revenge which occurred in France, in the reign of Charles V. [E] The anecdote has been frequently related, and is as follows:—A gentleman named Macaire, an officer of the king's body-guard, entertained, for some reason, a bitter hatred against another gentleman, named Aubry de Montdidier, his comrade in service. These two having met in the Forest of Bondi, near Paris, Macaire took an opportunity of treacherously murdering his brother-officer, and buried him in a ditch. Montdidier was unaccompanied at the moment, excepting by a dog (probably a wolf-hound), with which he had gone out, perhaps to hunt. It is not known whether the dog was muzzled, or from what other cause it permitted the deed to be accomplished without its interference. Be this as it might, the hound lay down on the grave of its master, and there remained till hunger compelled it to rise. It then went to the kitchen of one of Aubry de Montdidier's dearest friends, where it was welcomed warmly, and fed. As soon as its hunger was appeased the dog disappeared. For several days this coming and going was repeated, till at last the curiosity of those who saw its movements was excited, and it was resolved to follow the animal, Page 91 and see if anything could be learned in explanation of Montdidier's sudden disappearance. The dog was accordingly followed, and was seen to come to a pause on some newly-turned-up earth, where it set up the most mournful wailings and howlings. These cries were so touching, that passengers were attracted; and finally digging into the ground at the spot, they found there the body of Aubry de Montdidier. It was raised and conveyed to Paris, where it was soon afterwards interred in one of the city cemeteries.

The dog attached itself from this time forth to the friend, already mentioned, of its late master. While attending on him, it chanced several times to get a sight of Macaire, and on every occasion it sprang upon him, and would have strangled him had it not been taken off by force. This intensity of hate on the part of the animal awakened a suspicion that Macaire had had some share in Montdidier's murder, for his body showed him to have met a violent death. Charles V., on being informed of the circumstances, wished to satisfy himself of their truth. He caused Macaire and the dog to be brought before him, and beheld the animal again spring upon the object of its hatred. The king interrogated Macaire closely, but the latter would not admit that he had been in any way connected with Montdidier's murder.

Being strongly impressed by a conviction that the conduct of the dog was based on some guilty act of Macaire, the king ordered a combat to take place Page 92 between the officer and his dumb accuser, according to the practice in those days between human plaintiffs and defendants. This remarkable combat took place on the isle of Notre Dame at Paris, in presence of the whole court. The king allowed Macaire to have a strong club, as a defensive weapon; while, on the other hand, the only self-preservative means allowed to the dog consisted of an empty cask, into which it could retreat if hard pressed. The combatants appeared in the lists. The dog seemed perfectly aware of its situation and duty. For a short time it leapt actively round Macaire, and then, at one spring, it fastened itself upon his throat, in so firm a manner that he could not disentangle himself. He would have been strangled had he not cried for mercy, and avowed his crime. The dog was pulled from off him; but he was only liberated from its fangs to perish by the hands of the law. The fidelity of this dog has been celebrated in many a drama and poem, and there is a monument of him in basso relievo still to be seen in the castle of Montargis. The dog which attracted such celebrity has been usually called 'the dog of Montargis,' from the combat having taken place at the château of that name.

The strength of these dogs must have been very great. A nobleman informed me, that when he was a boy, and staying on a visit with the Knight of Kerry, two Irish wolf-dogs made their escape from the place in which they were confined, and pulled down and killed a horse, which was in an adjoining paddock.

Page 93 The following affecting anecdote of an Irish wolf-dog, called "the dog of Aughrim," affords a proof of the extraordinary fidelity of these animals to their masters, and puts to shame the vaunted superiority of many human brutes.

At the hard-fought battle of Aughrim, or Vidconnel, an Irish officer was accompanied by his wolf-hound. This gentleman was killed and stripped in the battle, but the dog remained by his body both by day and night. He fed upon some of the other bodies with the rest of the dogs, yet he would not allow them or anything else to touch that of his master. When all the other bodies were consumed, the other dogs departed, but this used to go in the night to the adjacent villages for food, and presently to return again to the place where his master's bones were only then left. This he continued to do from July, when the battle was fought, until the January following, when a soldier being quartered near, and going that way by chance, the dog, fearing he came to disturb his master's bones, flew upon the soldier, who, being surprised at the suddenness of the thing, unslung his carbine, he having been thrown on his back, and killed the noble animal. He expired with the same fidelity to the remains of his unfortunate master, as that master had shown devotion to the cause of his unhappy country.

In the "Irish Penny Journal" there is an interesting account of the Irish wolf-dog, from which the following anecdote is taken.

Page 94 In the mountainous parts of the county Tyrone, the inhabitants suffered much from the wolves, and gave from the public fund as much for the head of one of these animals, as they would now give for the capture of a notorious robber on the highway. There lived in those days an adventurer, who, alone and unassisted, made it his occupation to destroy these ravagers. The time for attacking them was in the night, and midnight was fixed upon for doing so, as that was their wonted time for leaving their lairs in search of food, when the country was at rest and all was still; then, issuing forth, they fell on their defenceless prey, and the carnage commenced. There was a species of dog for the purpose of hunting them, called the wolf-dog; the animal resembled a rough, stout, half-bred greyhound, but was much stronger. In the county Tyrone there was then a large space of ground enclosed by a high stone wall, having a gap at each of the two opposite extremities, and in this were secured the flocks of the surrounding farmers. But, secure as this fold was deemed, it was often entered by the wolves, and its inmates slaughtered. The neighbouring proprietors having heard of the noted wolf-hunter above mentioned, by name Rory Carragh, sent for him, and offered the usual reward, with some addition, if he would undertake to destroy the two remaining wolves that had committed such devastation. Carragh, undertaking the task, took with him two wolf-dogs, and a little boy twelve years of age, the only person who Page 95 would accompany him, and repaired at the approach of midnight to the fold in question. "Now," said Carragh to the boy, "as the two wolves usually enter the opposite extremities of the sheep-fold at the same time, I must leave you and one of the dogs to guard this one while I go the other. He steals with all the caution of a cat, nor will you hear him, but the dog will, and will give him the first fall. If, therefore, you are not active when he is down to rivet his neck to the ground with this spear, he will rise up and kill both you and the dog. So good night."

"I'll do what I can," said the little boy, as he took the spear from the wolf-hunter's hand.

The boy immediately threw open the gate of the fold, and took his seat in the inner part, close to the entrance, his faithful companion crouching at his side, and seeming perfectly aware of the dangerous business he was engaged in. The night was very dark and cold, and the poor little boy, being benumbed with the chilly air, was beginning to fall into a kind of sleep, when at that instant the dog, with a roar, leaped across, and laid his mortal enemy upon the earth. The boy was roused into double activity by the voice of his companion, and drove the spear through the wolf's neck as he had been directed, at which time Carragh appeared, bearing the head of the other.

This anecdote is taken from a biography of a Tyrone family, published in Belfast in 1829.

It is now time to attempt a description of this Page 96 celebrated dog, and here our difficulties commence. Some writers have affirmed that it was rough-coated, and had the appearance of a greyhound—

> "The greyhound! the great hound! the graceful of limb!
> Rough fellow! tall fellow! &c.;"

while others assert that it was of a mastiff-like appearance, and smooth, strong, and tall. All we can do is to bring forward the different evidence we have been able to collect, and then to let our readers judge for themselves.

In an old print of Patrick Sarsfield, Earl of Lucan, there are two wolf-dogs, which are represented as smooth, prick-eared, and with somewhat bushy tails. Lord Lucan distinguished himself in several engagements, and commanded the second troop of Irish Horse Guards, to which he was appointed by James II., and received his death wound, behaving most gallantly at the head of his countrymen, in 1693, when the allies, under William III., were defeated by Marshal Luxembourg at the battle of Landen. He was probably attended by his faithful wolf-dogs on that occasion, when he uttered those sublime words which no Irishman will ever forget—"Oh that this was for Ireland!" thus showing his love and affection for his native country as he was expiring in the arms of victory.

An old and amiable acquaintance, Mr. Aylmer Bourke Lambert, now, alas! no more, communicated an account of the wolf-hound to the Linnean Society, Page 97 which may be found in the third volume of their "Transactions." He had in his possession an old picture of one of these dogs, which, at the sale of his effects, was purchased by the Earl of Derby; the dog is represented as smooth-haired, with a somewhat wide forehead, and having no appearance of the greyhound, but more of that of the mastiff.

In February, 1841, Mr. Webber presented to the Royal Irish Academy an ancient stone, on which was carved a rude bas-relief, supposed to be the representation of a dog killing a wolf. Mr. Webber accompanied the present with a communication, to the effect that the stone was taken from the castle of Ardnaglass, in the barony of Tireragh, and county of Sligo, and was said to commemorate the destruction of the last wolf in Ireland. The current tradition in the place from whence it came was, that some years after it was supposed that the race of wolves was extinct, the flocks in the county of Leitrim were attacked by a wild animal, which turned out to be a wolf; that thereupon the chieftains of Leitrim applied to O'Dowd, the chieftain of Tireragh (who possessed a celebrated dog of the breed of the ancient Irish wolf-dog), to come and hunt the wolf. This application having been complied with by O'Dowd, there ensued a chase, which forms the subject of an ancient Irish legend, detailing the various districts through which it was pursued, until at length the wolf was overtaken and killed in a small wood of pine-trees, at the foot of one of the mountains of Page 98 Tireragh. The quarter of land on which the wolf was killed is to this day called *Carrow na Madhoo,* which means "the dog's quarter." In commemoration of the event, O'Dowd had a representation of it carved on stone, and placed in the wall of his baronial residence. It is difficult to form an opinion of the shape of a dog from so rude a representation, except that it appears to have had a wide forehead and pricked ears.

A gentleman, who in his youth saw one of these dogs, informs me that it was smooth, strong, and partaking somewhat of the character and appearance of a powerful Danish dog. This agrees with the account given of it by some writers, especially in "The Sportsman's Cabinet," a work more remarkable for the truth and fineness of its engravings, than for the matter contained in it. Buffon also forms much the same opinion. That great strength must be necessary to enable a dog to compete with a wolf, cannot be doubted, and perhaps there is no breed of the rough greyhound now known capable of competing with a wolf single-handed. Her Majesty has now in her possession one of the finest specimens of the Highland deerhound. He has great strength and height, is rough-coated, wide across the loins, and altogether a noble animal. Powerful, however as he is, it may be questioned whether such a dog would be a match for a wolf, which the Irish hounds undoubtedly were. This circumstance alone would lead us to suppose, that we must look to a different Page 99 breed than that of greyhounds as the antagonists of the wolf.

But it is time to turn to the other side of the question.

In a very agreeable, well-written article in the "Irish Penny Journal" of May, 1841, the author brings forward strong evidence to prove that the celebrated Irish wolf-dog resembled a greyhound in form. He will, I hope, allow me to quote some of his arguments, which show considerable research and historical information. He says:—

"Public opinion has long been divided respecting the precise appearance and form of this majestic animal, and so many different ideas have been conceived of him, that many persons have been induced to come to the conclusion that no particular breed of dogs was ever kept for wolf-hunting in Ireland, but that the appellation of 'wolf-dog' was bestowed upon any dog swift enough to overtake and powerful enough to contend with and overcome that formidable animal. While some hold this opinion, others suppose that though a particular breed was used, it was a sort of heavy mastiff-like dog, now extinct. It is the object of the present paper to show, that not only did Ireland possess a peculiar race of dogs, exclusively devoted to wolf-hunting, but that those dogs, instead of being of the mastiff kind, resembled the greyhound in form; and instead of being extinct are still to be met with, although they are very scarce. I myself was once in a Page 100 very gross error respecting this dog, for I conceived him to have been a mastiff, and implicitly believed that the dogs of Lord Altamont, described in the third volume of the Linnean 'Transactions' by Mr. Lambert, were the sole surviving representatives of the Irish wolf-dog. An able paper, read by Mr. Haffield about a year ago, before the Dublin Natural History Society, served to stagger me in my belief, and subsequent careful inquiry and research have completed my conversion. I proceed to lay before my readers the result of that inquiry, and I feel confident that no individual, after reading the evidence which I shall adduce, will continue to harbour a doubt respecting the true appearance and form of the ancient Irish wolf-dog.

"We are informed by several disjointed scraps of Celtic verse, that in the times of old, when Fionn Mac Cumhaill, popularly styled Finn Mac Cool, wielded the sceptre of power and justice, we possessed a prodigious and courageous dog, used for hunting the deer and wild boar, and also the wolf, which ravaged the folds and slaughtered the herds of our ancestors. We learn from the same source that these dogs were also frequently employed as auxiliaries in war, and that they were 'mighty in combat, their breasts like plates of brass, and greatly to be feared.' We might adduce the songs of Ossian, where the epithets 'hairy-footed,' 'white-breasted,' and 'bounding,' are singularly characteristic of some of the striking peculiarities of the dog in question, and strangely coincide with the descrip Page 101 tions furnished by other writers respecting him. Mac Pherson must, at all events, have been at the pains of considerable research if he actually forged the beautiful poems, which he put forth to the world under Ossian's name. The word 'Bran,' the name given to Fingal's noble hound, employed by others than Ossian, is Celtic, and signifies 'Mountain Torrent,' implying that impetuosity of course and headlong courage which the dog possessed. I have said that many assert the Irish wolf-dog to be no longer in existence. I have ventured a denial of this, and refer to the wolf-dog or deer-dog of the Highlands of Scotland, as his actual and faithful living representative. Perhaps I am wrong in saying representative. I hold that the Irish wolf-dog and the Highland deer-dog are one and the same, and I now proceed to cite a few authorities in support of my position.

"The Venerable Bede, as well as the Scotch historian John Major, informs us that Scotland was originally peopled from Ireland under the conduct of Renda, and that one half of Scotland spoke the Irish language as their mother-tongue. Many persons, also, are doubtless aware that, even at this present time, the Gaelic and Erse are so much alike, that a Connaught man finds no difficulty in comprehending and conversing with a Highlander. Scotland also was called by the early writers Scotia Minor, and Ireland, Scotia Major. The colonization, therefore, of Scotland from Ireland admits of little doubt. As the Irish wolf-dog Page 102 was at that time in the enjoyment of his most extended fame, it was not to be expected that the colonists would omit taking with them such a fine description of dog, and which would prove so useful to them in a newly established settlement, and that, too, at a period when hunting was not merely an amusement, but one of their main occupations, and also their main source of subsistence. The Irish wolf-dog was thus carried into Scotland, and became the Highland or Scottish wolf-dog, changing in process of time his name with his country; and when wolves disappeared from the land, his occupation was that of deer-hunting, and thus his present name.

"In Ireland the wolves were in existence longer than in Scotland, but as soon as wolves ceased to exist in the former country, the dogs were suffered to become extinct also, while in Scotland there was still abundant employment for them after the days of wolf-hunting were over—the deer still remained; and useful as they had been as wolf-dogs, they proved themselves, if possible, still more so as deer-hounds.

"That the Irish wolf-dog was a tall, rough greyhound, similar in every respect to the Highland dog of the present day (of which an engraving is given) cannot be doubted from the following authorities. Strabo mentions a tall greyhound in use among the Pictish and Celtic nations, which he states was held in high esteem by our ancestors, and was even imported into Gaul for the purposes of the chase. Page 103 Campion expressly speaks of the Irish wolf-dog as a 'greyhound of great bone and limb.' Silaus calls it also a greyhound, and asserts that it was imported into Ireland by the Belgæ, and is the same with the renowned Belgic dog of antiquity, and that it was, during the days of Roman grandeur, brought to Rome for the combats of the Amphitheatre. Pliny relates a combat in which the Irish wolf-dog took a part: he calls them 'Canes Graii Hibernici,' and describes them as much taller than the mastiff. Holinshed, in speaking of the Irish, says, 'They are not without wolves, and greyhounds to hunt them.' Evelyn, speaking of the bear-garden, says, 'The bull-dogs did exceeding well, but the Irish wolf-dog exceeded; which was a tall greyhound, a stately creature, and beat a cruel mastiff.'

"Llewellyn, prince of Wales, was presented by King John with a specimen of this kind of dog. These animals were in those days permitted to be kept only by princes and chiefs; and in the Welsh laws of the ninth century we find heavy penalties laid down for the maiming or injuring of the Irish greyhound, or, as it was styled in the code alluded to, 'Canis Graius Hibernicus;' and a value was set on them, equal to more than double that set on the ordinary greyhound.

"Moryson, secretary to Lord-deputy Mountjoy, says, 'The Irishmen and greyhounds are of great stature.' Lombard remarks, that the finest hunting dogs in Europe were produced in Ireland: 'Greyhounds useful to take the stag, wild boar, or wolf.' Pennant Page 104 describes these dogs as scarce, and as being led to the chase in leather slips or thongs, and calls them 'the Irish greyhound.' Bay mentions him as the greatest dog he had ever seen. Buffon says, he saw an Irish greyhound, which measured five feet in height when in a sitting posture, and says that all other sorts of greyhounds are descended from him, and that in Scotland it is called the Highland greyhound: that it is very large, deep-chested, and covered with long rough hair.

"Scottish noblemen were not always content with such specimens of this dog as their own country produced, but frequently sent for them to Ireland, conceiving, doubtless, that they would be found better and purer in their native land. The following is a copy of a letter addressed by Deputy Falkland to the Earl of Cork, in 1623:—

'My Lord,

'I have lately received letters from my Lord Duke of Buccleuch and others of my noble friends, who have entreated me to send them some greyhound dogs and bitches, out of this kingdom, of the largest sort, which I perceive they intend to present unto divers princes and other noble persons; and if you can possibly, let them be white, which is the colour most in request here. Expecting your answer by the bearer, I commit you to the protection of the Almighty, and am your Lordship's attached friend,

'Falkland.'

Page 105 "Smith, in his 'History of Waterford,' says, 'the Irish greyhound is nearly extinct: it is much taller than a mastiff, but more like a greyhound, and for size, strength, and shape, cannot be equalled. Roderick, king of Connaught, was obliged to furnish hawks and greyhounds to Henry II. Sir Thomas Rue obtained great favour from the Great Mogul in 1615, for a brace of Irish greyhounds presented by him. Henry VIII. presented the Marquis of Dessarages, a Spanish grandee, with two goshawks and four Irish greyhounds.'

"Perhaps sufficient evidence has now been adduced to demonstrate the identity of the Irish wolf-dog with the Highland deerhound. I may, however, in conclusion, give an extract from the excellent paper of Mr. Haffield, already alluded to, as having been read before the Dublin Natural History Society, and which was received by that gentleman from Sir William Betham, Ulster King-at-Arms, an authority of very high importance on any subject connected with Irish antiquities. Sir William says,—'From the mention of the wolf-dogs in the old Irish poems and stories, and also from what I have heard from a very old person, long since dead, of his having seen them at 'The Neale,' in the county of Mayo, the seat of Sir John Browne, ancestor to Lord Kilmaine, I have no doubt they were a gigantic greyhound. My departed friend described them as being very gentle, and says that Sir John Browne allowed them to come Page 106 into his dining-room, where they put their heads over the shoulders of those who sat at table. They were not smooth-skinned, like our greyhounds, but rough and curly-haired. The Irish poets call the wolf-dog 'Cu,' and the common greyhound 'Gayer;' a marked distinction, the word 'Cu' signifying a champion.'

"The colour of these dogs varies, but the most esteemed are dark iron-grey, with white breast. They are, however, to be found of a yellowish or sandy hue, brindled, or even white. In former times, as will be seen from Lord Falkland's letter quoted above, this latter colour was by many preferred. It is described as a stately, majestic animal, extremely good-tempered and quiet in his disposition, unless when irritated or excited, when he becomes furious; and is, in consequence of his tremendous strength, a truly formidable animal."

Goldsmith asserts that he had seen a dozen of these dogs, and informs us "that the largest was about four feet high, or as tall as a calf of a year old. They are generally of a white or cinnamon colour, and more robust than the greyhound—their aspect mild, and their disposition gentle and peaceable. It is said that their strength is so great, that in combat the mastiff or bull-dog is far from equal to them. They commonly seize their antagonists by the back and shake them to death. These dogs were never serviceable for hunting, either the stag, the fox, or the hare. Page 107 Their chief utility was in hunting wolves, and to this breed may be attributed the final extirpation of those ferocious animals in England and Wales in early times in the woody districts."

Having thus given these different accounts of the Irish wolf-dog, I may add that some persons are of opinion that there were two kinds of them—one partaking of the shape and disposition of the mastiff, and the other of the Highland deer-hound. It is not improbable that a noble cross of dogs might have been made from these two sorts.

At all events I have fairly stated the whole of the information I have been able to obtain respecting these dogs, and my readers must form their own opinions. The following anecdote, recently communicated to me, is given in the words of the writer:—

"Two whelps were made a present to my brother by Harvey Combe, of a breed between the old Irish wolf-dog and the bloodhound. My brother gave them to Robert Evatt, of Mount Louise, county Monaghan. One died young, but the other grew to be a very noble animal indeed. Unfortunately he took to chasing sheep, and became an incorrigible destroyer of that inoffensive but valuable stock. Evatt found he could not afford to keep such a marauder, and as he was going to Dublin he took up the sheep-killer, in order to present him to the Zoological Society as a fine specimen of the breed. His servant was holding him at the door of the hotel when a gig drove up, and Page 108 the gentleman alighted. The dog sprung from the servant's hold, and jumping into the gig with one bound, seized the mat at the bottom of the gig, which was made of sheepskin, and with another bound made away with his woolly prize, and was brought back with difficulty, after a long and fatiguing pursuit."

This is one of the most desperate cases of sheep-hunting in dogs I ever met with. It is said, that this propensity may be got rid of by tying a cord covered with wool to the dog's lower jaw, so that the wool may be kept in the mouth.

I should mention, that in a manuscript of Froissart in the British Museum, which is highly illuminated, there is a representation of the grand entrance of Queen Isabel of England into Paris, in the year 1324. She is attended by a noble greyhound, who has a flag, *powdered* with fleurs-de-lys, bound to his neck.

Greyhounds were a favourite species of dog in the middle ages. In the ancient pipe-rolls, payments are frequently made in greyhounds. In Hawes' "Pastime of Pleasure," (written in the time of Henry VII.) Fame is attended by two greyhounds, on whose golden collars, "Grace" and "Governaunce" are inscribed in diamond letters.

In the pictures of Rubens, Snyders, and other old masters, some of the powerful dogs there represented would appear to be a breed between the greyhound and mastiff. Nothing can exceed the majestic and Page 109 commanding appearance of these dogs, and such a breed would be most likely to produce the sort of animal most capable of contending with the wolf.

The Irish wolf-dogs were formerly placed as the supporters of the arms of the ancient Monarchs of Ireland. They were collared *or*, with the motto,

"Gentle when stroked—fierce when provoked."

Mr. Scrope, in his agreeable book on deer-stalking in Scotland, has communicated an account from Mr. Macneill, of Colonsay, of the Highland deer-hound, in which are some interesting remarks relative to the Irish wolf-dog, and from which I shall make a few extracts.

In making these extracts, it is impossible not to be struck with a remark in the work referred to, that from modern writers we learn nothing further respecting the Irish wolf-dog, than that such a race of dogs at one time existed in Ireland, that they were of a gigantic size, and that they are now extinct.

One great obstacle in the way of investigating the history of this dog has arisen from the different appellations given to it, according to the fancy of the natives in different parts of the country, such as Irish wolf-dog, Irish greyhound, Highland deer-hound, and Scotch greyhound, and this circumstance may have produced the confusion in fixing its identity.

In the fourth century a number of dogs, of a great Page 110 size, were sent in iron cages from Ireland to Rome, and it is not improbable that the dogs so sent were greyhounds, particularly as we learn from the authority of Evelyn and others, that the Irish wolf-dog was used for the fights of the bear-garden. "Greyhound" probably means a "great hound."

Holinshed, in his "Description of Ireland and the Irish," written in 1586, has the following notice:—"They are not without wolves, and greyhounds to hunt them, bigger of bone and limb than a colt;" and in a frontispiece to Sir James Ware's "History of Ireland," an allegorical representation is given of a passage from the Venerable Bede, in which two dogs are introduced, bearing a strong resemblance to that given by Gesner, in his "History of Quadrupeds," published in 1560.

The term *Irish* is applied to Highland dogs, as everything Celtic (not excepting the language) was designated in England; probably in consequence of Ireland being, at that period, better known to the English than Scotland. This is, perhaps, a proof of the similarity of the Irish and Scotch deer-hounds.

Of the courage of the ancient deer-hound there can be little doubt, from the nature of the game for which he was used. If any proof were wanting, an incident mentioned by Evelyn in his Diary, in 1670, when present at a bull-fight in the bear-garden, is conclusive. He says, "The bulls (meaning the bull-dogs) did exceeding well, but the Irish wolf-dog ex Page 111 ceeded, which was a tall greyhound, a stately creature, indeed, who beat a cruel mastiff."

Here, perhaps, is a proof that the Irish wolf-dog was a greyhound; and there can be little doubt that it is the same dog we find mentioned under the name of the Irish greyhound.

Buffon remarks that "the Irish greyhounds are of a very ancient race. They were called by the ancients, dogs of Epirus, and Albanian dogs. Pliny gives an account of a combat between one of these dogs, first with a lion, and then with an elephant. In France they are so rare, that I never saw above one of them, which appeared, when sitting, to be about five feet high. He was totally white, and of a mild and peaceable disposition."

The following description of these dogs, translated from a Celtic poem, is probably an accurate one:—

"An eye of sloe, with ear not low,
With horse's breast, with depth of chest,
With breadth of loin, and curve in groin
And nape set far behind the head—
Such were the dogs that Fingal bred."

It is probable that even in Scotland very few of the pure breed of dogs are left, but those which are show a surprising combination of speed, strength, size, endurance, courage, sagacity, docility, and it may be added, dignity. The purest specimens of the deer-hound now to be met with are supposed to be those belonging to Captain M'Neill of Colonsay, two of Page 112 them being called Buskar and Bran. And here let me give an extract from an interesting and graphic account, published by Mr. Scrope, of the performance of these dogs in the chase of a stag. Let us fancy a party assembled over-night in a Highland glen, consisting of sportsmen, deer-stalkers, a piper and two deer-hounds, cooking their supper, and concluding it with the never-failing accompaniment of whisky-toddy. Let us fancy them reposing on a couch of dried fern and heather, and being awoke in the morning with the lively air of "Hey, Johnny Cope." While their breakfast is preparing, they wash and refresh themselves at a pure mountain stream, and are soon ready to issue forth with Buskar and Bran. The party proceeds up a rocky glen, where the stalker sees a stag about a mile off. He immediately prostrates himself on the ground, and in a second the rest follow his example. We will not follow all the different manœuvres of the deer-stalker and his followers, but bring them at once near the unconscious stag. After performing a very considerable circuit, moving sometimes forwards and sometimes backwards, the party at length arrive at the back of a hillock, on the opposite side of which the stalker said, in a whisper, the deer was lying, and that he was not distant a hundred yards. The whole party immediately moved forward in silent and breathless expectation, with the dogs in front straining in the slips. On reaching the top of the hillock, a full view of the noble stag pre Page 113 sented itself, who, having heard the footsteps, had sprung on his legs, and was staring at his enemies, at the distance of about sixty yards.

"The dogs were slipped; a general halloo burst from us all, and the stag, wheeling round, set off at full speed, with Buskar and Bran straining after him.

"The brown figure of the deer, with his noble antlers laid back, contrasted with the light colour of the dogs stretching along the dark heath, presented one of the most exciting scenes that it is possible to imagine.

"The deer's first attempt was to gain some rising ground to the left of the spot where we stood, and rather behind us, but, being closely pursued by the dogs, he soon found that his only safety was in speed; and (as a deer does not run well up-hill, nor like a roe, straight down hill) on the dogs approaching him, he turned, and almost retraced his footsteps, taking, however, a steeper line of descent than the one by which he ascended. Here the chase became most interesting—the dogs pressed him hard, and the deer getting confused, found himself suddenly on the brink of a small precipice of about fourteen feet in height, from the bottom of which there sloped a rugged mass of stones. He paused for a moment, as if afraid to take the leap, but the dogs were so close that he had no alternative.

"At this time the party were not above one Page 114 hundred and fifty yards distant, and most anxiously waited the result, fearing, from the ruggedness of the ground below, that the deer would not survive the leap. They were, however, soon relieved from their anxiety, for though he took the leap, he did so more cunningly than gallantly, dropping himself in the most singular manner, so that his hind legs first reached the broken rocks below; nor were the dogs long in following him. Buskar sprang first, and, extraordinary to relate, did not lose his legs. Bran followed, and, on reaching the ground, performed a complete somerset. He soon, however, recovered his legs, and the chase was continued in an oblique direction down the side of a most rugged and rocky brae, the deer, apparently more fresh and nimble than ever, jumping through the rocks like a goat, and the dogs well up, though occasionally receiving the most fearful falls.

"From the high position in which we were placed, the chase was visible for nearly half a mile. When some rising ground intercepted our view, we made with all speed for a higher point, and, on reaching it, we could perceive that the dogs, having got upon smooth ground, had gained on the deer, who was still going at speed, and were close up with him. Bran was then leading, and in a few seconds was at his heels, and immediately seized his hock with such violence of grasp, as seemed in a great measure to paralyse the limb, for the deer's speed was immediately checked. Buskar was not far behind, for soon afterwards Page 115 passing Bran, he seized the deer by the neck. Notwithstanding the weight of the two dogs which were hanging to him, having the assistance of the slope of the ground, he continued dragging them along at a most extraordinary rate (in defiance of their utmost exertions to detain him), and succeeded more than once in kicking Bran off. But he became at length exhausted—the dogs succeeded in pulling him down; and though he made several attempts to rise, he never completely regained his legs.

"On coming up, we found him perfectly dead, with the joints of both his forelegs dislocated at the knee, his throat perforated, and his chest and flanks much lacerated.

"As the ground was perfectly smooth for a considerable distance round the place where he fell, and not in any degree swampy, it is difficult to account for the dislocation of his knees, unless it happened during his struggles to rise. Buskar was perfectly exhausted, and had lain down, shaking from head to foot much like a broken-down horse; but on our approaching the deer he rose, walked round him with a determined growl, and would scarcely permit us to get near him. He had not, however, received any cut or injury, while Bran showed several bruises, nearly a square inch having been taken off the front of his fore-leg, so that the bone was visible, and a piece of burnt heather had passed quite through his foot.

"Nothing could exceed the determined courage Page 116 displayed by both dogs, particularly by Buskar, throughout the chase, and especially in preserving his hold, though dragged by the deer in a most violent manner."

It is hoped that this account of the high spirit and perseverance of the Scotch deer-hound will not be found uninteresting. This noble creature was the pride and companion of our ancestors, and for a long period in the history of this country, particularly in Ireland, the only dog used in the sports of the field. When we consider the great courage, combined with the most perfect gentleness of this animal, his gigantic, picturesque, and graceful form, it must be a subject of regret that the breed is likely to become extinct. Where shall we find dogs possessing such a combination of fine and noble qualities?

The following anecdote, which with the accompanying fine engraving is taken from the New Sporting Magazine for January 1839, presents a striking example of the same kind:—

"The incident which the artist has made the subject for our embellishment occurred with Lord Ossulston's stag-hounds, on Tuesday, the 1st of May, when the stag, after a fast run of an hour, jumped over a precipice, and broke his neck. The hounds were, at this time, close to his haunches, and a couple and a half of the leading dogs went over with the stag. Two of the hounds were so hurt that they could not move, and Page 117 the third was found by the greencoat first up, lying on the dead deer."

I am indebted to that clever and intelligent authoress, Mrs. S. Carter Hall, for her recollections of an Irish wolf-dog and his master, which I cannot do better than give in her own words:—

"When I was a child, I had a very close friendship with a genuine old wolf-dog, Bruno by name. He was the property of an old friend of my grandmother's, who claimed descent from the Irish kings. His name was O'Toole. His manners were the most courtly you can imagine; as they might well be, for he had spent much time and fortune at the French court, when Marie Antoinette was in her prime and beauty. His visits were my jubilees—there was the kind, dignified old gentleman, who told me tales—there was his tall, gaunt dog, grey with age, and yet with me full of play; and there were two rough terriers, whom Bruno kept in admirable order. He managed the little one by simply placing his paw upon it when it was too frisky; but Vixen, the large one, like many ladies, had a will of her own, and entertained some idea of being mistress. Bruno would bear a good deal from her, giving, however, now and then, a low deep growl; but when provoked too much, he would quietly lift the dog off the ground by the strength of his jaws (his teeth were gone), stand with her in his mouth at the doors until they were opened, and then deposit her, half Page 118 strangled as she was, in a nettle-bed some distance from the house. The dog's discrimination was curious. If Vixen was thrown upon him, or if we forced her to insult him, he never punished her; but if she of her own accord teazed him more than his patience could bear, the punishment was certain to follow.

"O'Toole and his dogs always occupied the same room, the terriers being on the bed with their master. No entreaty, however, ever induced Bruno to sleep on anything softer than stone. He would remove the hearth-rug and lay on the marble. His master used to instance the dog's disdain of luxury as a mark of his noble nature.

"I should not omit to tell you, as characteristic of my old friend, that O'Toole was proud, and never would submit to be called 'Mr.' Meeting, one day, Lord Arne in Dame Street, Dublin, while the old man was followed by his three wolf-dogs, of which Bruno was the last, the young nobleman, who had also his followers in the shape of 'Parliament men,' said to the descendant of Irish kings, nodding to him familiarly at the same time, 'How do you do, *Mr.* O'Toole?' The old man paused, drew himself up, lifted his hat, made his courtly bow, and answered, 'O'Toole salutes Arne.' I can recall nothing more picturesque than that majestic old gentleman and his dog, both remnants of a bygone age. Bruno was rough, but not long-coated, very grave, observant, enduring every one, very fond of children, playing with them Page 119 gently, but only crouching and fawning on his master; 'and that,' O'Toole would say, 'is a proof of my royal blood.' I could fill a volume with memoirs of that fine old man. He was more than six feet in height, and his dog always sat with his head on his master's knee."

This is altogether a pretty and interesting picture.

The sagacity of this fine breed is well illustrated in what follows:—

A gentleman walking along the road on Kingston Hill, accompanied by a friend and a noble deer-hound, which was also a retriever, threw his glove into a ditch; and having walked on for a mile, sent his dog back for it. After waiting a considerable time, and the dog not returning, they retraced their steps. Hearing loud cries in the distance, they hastened on, and at last saw the dog dragging a boy by his coat towards them. On questioning the boy, it appeared that he had picked up the glove and put it into his pocket. The sagacious animal had no other means of conveying it to his master than by compelling the boy to accompany him.

The following anecdotes are from Capt. Thomas Brown's now scarce work, "Biographical Sketches and Anecdotes of Dogs." He says: —

"Sir Walter Scott has most obligingly furnished me with the following anecdotes of his celebrated dog Maida: Page 120 —

"I was once riding over a field on which the reapers were at work, the stooks being placed behind them, as is usual. Maida having found a hare, began to chase her, to the great amusement of the spectators, as the hare turned very often and very swiftly among the stooks. At length, being hard pressed, she fairly bolted into one of them. Maida went in headlong after her, and the stook began to be much agitated in various directions. At length the sheaves tumbled down; and the hare and the dog, terrified alike at their overthrow, ran different ways, to the great amusement of the spectators."

"Among several peculiarities which Maida possessed, one was a strong aversion to a certain class of artists, arising from the frequent restraints he was subjected to in having his portrait taken, on account of his majestic appearance. The instant he saw a pencil and paper produced he prepared to beat a retreat; and, if forced to remain, he exhibited the strongest marks of displeasure."

Ranaldson Macdonell, Esq. of Glengarry, has most kindly furnished the following interesting notices and anecdotes of the Scottish Highland greyhound: —

"Not many years since one of Glengarry's tenants, who had some business with his chief, happened to arrive at Glengarry House at rather an early hour in the morning. A deer-hound perceiving this person sauntering about before the domestics were astir, Page 121 walked quietly up to him, took him gently by the wrist with his teeth, and proceeded to lead him off the ground. The man, finding him forbearing, attempted resistance; but the dog, instantly seizing his wrist with redoubled pressure, soon convinced him that his attempt was in vain. Thus admonished, the man took the hint, and quietly yielded to his canine conductor, who, without farther injury, led him to the outside of the gate, and then left him. The whole of the dogs at Glengarry House were allowed to go at liberty at all times.

"The Highland greyhounds, or deer-hounds as they are called in the Highlands, have a great antipathy to the sheep-dogs, and never fail to attack them whenever an opportunity offers. A shepherd, whose colley had frequently been attacked by the deer-dogs of Glengarry singly, and always succeeded in beating them off on such occasions, was one day assailed by them in a body; and his life would have been in considerable danger, but for one of the keepers, who happened to pass at the time, and called them off.

"The following circumstance will prove the exquisite sense of smell possessed by the deer-hound. One of this breed, named Bran, when held in the leash, followed the track of a wounded stag, and that in most unfavourable rainy weather, for three successive days, at the end of which time the game was shot. He was wounded first within nine miles of Inver Page 122 garry House, and was traced that night to the estate of Glenmoriston. At dusk in the evening the deer-stalkers placed a stone on each side of the last fresh print of his hoof, and another over it; and this they did each night following. On the succeeding morning they removed the upper stone, when the dog recovered the scent, and the deer was that day traced over a great part of Glenmoriston's ground. On the third day he was re-traced to the lands of Glengarry, and there shot.

"My present dog, Comhstri, to great courage unites the quality of a gentle disposition, with much fidelity and attachment. Though not so large as some of his kindred, he is nevertheless as high-spirited and determined as any of his race, which the following circumstance will testify: 'About three years ago, a deer from the wood of Derrygarbh, whose previous hurts had been healed, came out of Glengarry's pass, who wounded it severely in the body with a rifle bullet. The deer-hounds were immediately laid on the blood-track. The stag was started in the course of a few minutes; the dogs were instantly slipped, and the fine animal ran to bay in a deep pool of water, below a cascade, on the Garyquulach burn. Comhstri immediately plunged in, and seized the stag by the throat; both went under water, surrounded with the white foam, slightly tinged with the deer's blood. The dog soon came to the surface to recover his breath; Page 123 and before the other could do so, Comhstri dived, and again seized him by the throat. The stag was soon after taken out of the pool dead.

"Comhstri's colour is grey, with a white chest; but we have had them of different colours at Glengarry, such as pure white, black, brindled, and sand-colour.

"When the Highlanders dream of a *black* dog, it is interpreted to mean one of the clan of Macdonell; but if of a deer-hound, it denotes a chief, or one of the principal persons of that clan."

That the Scottish dogs were much prized in England from the earliest times, the following interesting account, taken from Holinshed's Chronicles, 'Historie of Scotland,' p. 71, printed in 1586, will show. "And shortlie after the return of these ambassadors into their countrie, divers young gentlemen of the Pictish nobilitie repaired unto King Crathlint, to hunt and make merie with him; but when they should depart homewards, perceiving that the Scotish dogs did farre excell theirs, both in fairnesse, swiftnesse, hardinesse, and also in long standing up and holding out, they got diverse both dogs and bitches of the best kinds for breed to be given them by the Scotish Lords; and yet not so contented, they stole one belonging to the king from his keeper, being more esteemed of him than all the others which he had about him. The master of the leash being informed hereof, pursued after them which had Page 124 stollen that dog, thinking indeed to have taken him from them; but they not willing to part with him, fell at altercation, and in the end chanced to strike the maister of the leash through with their horsespeares that he died presentlie: whereupon noise and crie being raised in the countrie by his servants, diverse of the Scots, as they were going home from hunting, returned, and, falling upon the Picts to revenge the death of their fellow, there ensued a shrewd bickering betwixt them, so that of the Scots there died three score gentlemen, besides a great number of the commons, not one of them understanding (till all was done) what the matter meant. Of the Picts there were about an hundred slaine. This circumstance led to a bloody war betwixt the two nations."

The following interesting anecdote, related by Mr. Carr in his "Stranger in Ireland," there can be no doubt, I think, refers to the Irish wolf-dog. Mr. Carr says, that while on his journey to Ireland he "wandered to a little church, which owed its elevation to the following circumstance. Llewelyn the Great, who resided near the base of

Snowdon, had a beautiful dog named Gelert, which had been presented to him by King John in 1205. One day, in consequence of the faithful animal, which at night always 'sentinelled his master's bed,' not making his appearance in the chase, Llewelyn returned home very angry, and met the dog, covered with blood, at the door of the chamber of his Page 125 child. Upon entering it, he found the bed overturned, and the coverlet stained with gore. He called to his boy; but receiving no answer, he rashly concluded that he had been killed by Gelert, and in his anguish instantly thrust his sword through the poor animal's body. The Hon. Robert Spencer has beautifully told the remainder of the story.

'His suppliant looks, as prone he fell,
No pity could impart;
But still his Gelert's dying yell
Passed heavy on his heart.

Arous'd by Gelert's dying yell,
Some slumb'rer waken'd nigh:
What words the parent's joy could tell,
To hear his infant's cry?

Nor scathe had he, nor harm, nor dread:
But the same couch beneath,
Lay a gaunt wolf all torn and dead,
Tremendous still in death.

Ah! what was then Llewelyn's pain?
For now the truth was clear:—
His gallant hound the wolf had slain,
To save Llewelyn's heir.' [F]

In order to mitigate his offence, Llewelyn built this chapel, and raised a tomb to poor Gelert; and the spot to this day is called *Beth-Gelert*, or the Grave of Gelert."

I should not omit to mention, that in Mr. Windle's Page 126 account of Cork, Kerry, &c., there is the following notice of the wolf and Irish wolf-dog.

"The last wolf seen in Ireland was killed in the neighbourhood of Annascuit, near Dingle, in 1710. The place is still known by the name of the Wolf's Step. The Irish called the wolf-dog *Sagh cliun*; and old Campion, speaking of the Irish, says, They are not without wolves, and greyhounds to hunt them bigger of bone and limne than a colt."

This noble animal is also described as "similar in shape to a greyhound, larger than a mastiff, and tractable as a spaniel."

The following fact will serve to prove that the deer-hound is possessed of a fine sense of smelling, a circumstance which has been doubted by many persons.

The head keeper of Richmond Park is possessed of a famous old deer-hound bitch, remarkable for her sagacity, and for having taken five bucks in one day. After a battue in the Park in the winter of 1845, he directed one of the under-keepers to examine the ground carefully, which had been shot over the day before. He was accompanied by the old dog, who was to act as retriever. She came to a point in one of the covers, as was her custom when she seemed to find a rabbit; but the keeper, finding that it was a hare, called her off. After going some distance, the dog went back and pointed the hare a second time. The keeper put her up, and then found that she had been wounded, having had her hind leg broken. Here the fine sense Page 127 of smelling was the more remarkable, as this old dog will not look at a hare, nor indeed can she be induced to run after one.

One of her progeny ran a wounded buck into the large pond in the Park, swam after it, killed it in the water, and then seizing it by the foot, swam with it to the shore.

Having now given my reader all the information I can gather on this dog of bygone times, I will gratify him with a letter I have received from a lady whose name is dear to Ireland, and highly placed in the ranks of English Literature:—

"Dear Sir,

"I am much flattered by your compliment to my national erudition, a very scanty stock in my best of times, and now nearly used up, in 'furnishing forth' the pages of many an idle tale, worked out in the 'Irish Interest,' as the mouse nibbled at the lion's net,—the same presumption, if not with the same results! However, I will rub up my old '*Shannos*,' as Elizabeth said of her Latin, and endeavour to recollect the little I have ever known on the subject of the Irish wolf-dog.

"Natural history is too much a matter of fact to have ever interested the poetic temperament of the Irish; Schools of Poetry, Heraldry, and Music, were opened (says the Irish historians), 'time immemorial.' St. Patrick found the Academies of Lismore and Page 128 Armagh in a flourishing condition, when he arrived on his great mission; and the more modern College of Clonard (founded in the fifth century by Bishop Finnan), had a great reputation for its learning and learned professors. But it does not appear that there was any Chair of Natural History or Philosophy in these scholastic Seminaries. Their Transactions recorded the miracles of saints rather than the miracles of nature. And had some daring Cuvier, or enterprising Lyell or Murchison, opened those spacious cabinets, once

'In the deep bosom of the ocean buried,'

or entombed in mountain layers for unnumbered ages, the Druid priests would probably have immolated the daring naturalist under his highest oak. Is it quite sure that the Prior of Armagh, or the founder of the Royal Academy of Clonard, the good Saint Finnan himself, would have served them much better? Certain, however, it is, that the Druids, Bards, Filiahs, Senachies and Saints of Ireland, who left such mighty reputations behind them for learning, have not dropped one word on the subject of the natural history of their 'Isle of Song;' and though they may have dabbled a little in that prosaic pursuit, they probably soon discovered its perilous tendency, and sang with the last and most charming of Irish Bards,—

'No, Science, to you
We have long bade a last and careless adieu.'

Page 129 "Nearly two thousand years after the foundation of the most learned Academies of Ireland, a pretty little Zoological Garden was opened in the capital of the country; but no living type of the

Irish wolf-dog is to be found there, nor were any 'fossil remains' of the noble animal discovered in the Wicklow Mines, [G] which were worked some fifty years back, but which, for want of capital or perseverance, only furnished a few Cronobane halfpence, and materials for a musical farce to one of the most delightful farcical Irish writers of his time; [H] for in Ireland,

'Tout finis par un chanson,'

(as Figaro had it of the France of his age,) when worse results do not follow disappointment.

"The Irish wolf-dog, therefore, it may be asserted, belongs to the poetical traditions of Ireland, or to its remote Milesian histories. 'Gomer, the eldest son of Japhet, and others, the immediate posterity of Noah, after the dispersion of mankind at Babel, ventured (it is said), to 'commit themselves by ships upon the sea,' to search out the unknown corners of the world, and thus found out a western land called Ireland.'—(Dr. Warner.)

Page 130 "It is probable they were the first to disturb its tranquillity by the introduction of wolves, a fragment of the menagerie of the Ark; for all noxious and destructive animals and reptiles were brought into Ireland by her invaders. The soil and clime of the 'woody Morven,' however, though not genial to their naturalisation, was long a prey to one of the most ferocious animals imported by foreign aggression to increase and multiply. Ireland swarmed with wolves, and its colonists and aborigines would in time have alike shared the fate of 'little Red Riding Hood;' when, lo! up started the noble *Canis familiaris Hibernicus*, which, greatly improved by a cross with the wolf itself, was found everywhere in fierce antagonism with foreign ferocity; and for his eminent services was not only speedily adopted by patriot kings and heroes, as part of their courtly and warlike parade, but sung by bards and immortalised by poets, as worthy of such illustrious companionship. It is thus Bran, the famous and beloved hound of Fingal, has become as immortal as his master; and a track is still shown on a mountain in Tyrone, near New Town Stuart, called 'The Track of the Foot of Bran, the Hound of Fionne Mac Cumhall.' So much for poetry and tradition. Modern naturalists, however, in their animal biography and prosaic view of things, have assigned the introduction of the wolf-dog in Ireland to the Danes, who brought it over in their first invasion; and its resemblance to '*Le gros Danois*' of Buffon favours the Page 131 supposition. 'When Ireland swarmed with wolves,' says Pennant, 'these dogs were confined to the chase; but as soon as these animals were extirpated, the number of the dogs decreased, and from that period were kept chiefly for state.' Goldsmith mentions having only seen in his time in Ireland one Irish wolf-hound that was four feet high. And though the father of the late Marquis of Sligo endeavoured to preserve the breed, his kennels in latter years exhibited but a scanty specimen. These majestic and beautiful animals are now, I believe, quite extinct in Ireland, where their scarcity is accounted for by Mr. Pennant as 'the consequence of the late King of Poland having procured from thence by his agents as many as could be purchased.' The last notice taken of the Irish wolf-dog in fictitious narrative may, I believe, be found in one of my own national novels, 'O'Donnel,' where the hero and his hound are first introduced to the reader together. I borrowed the picture, as I gave it, from living originals, which in my earliest youth struck forcibly on my imagination, in the person of the celebrated Archibald Hamilton Rowan, accompanied by his Irish hound Bran!

"This is all I know or can recollect of my noble and beautiful compatriot; but I remember that when some writer in 'Fraser's Magazine' styled me 'that Irish she wolf-dog,' I felt complimented by the epithet, since to attack the enemies of Ireland, and to worry Page 132 when they could not destroy them, was the peculiar attribute of the species.

"I have the honour to be, dear Sir,

"Most truly yours,

"Sydney Morgan."

"*William Street, Albert Gate.*"

Page 133

THE NEWFOUNDLAND DOG.

"Nor will it less delight th' attentive sage,
T' observe that instinct which unerring guides
The brutal race, which mimics reason's lore,
And oft transcends.
*   *   *   *
The dog, whom nothing can mislead,

Must be a dog of parts indeed.
Is often wiser than his master."—Sommerville.

This noble dog may be justly styled the friend and guardian of his master. I had some doubts in making out my list of dogs, whether he ought not to take precedence of all others; but, after duly weighing the Page 134 matter in my own mind, I have given the palm to the Irish wolf-hound, and the honest Newfoundland immediately follows him. I not only think that this precedence will gratify some of my friends in Ireland, who have called upon me to do justice to one of their favourite and national emblems, but it is, perhaps, due in strict justice to an animal who proved himself so great a benefactor to his native country. There is, moreover, such a degree of romance attached to the recollection of his fine qualities and imposing appearance, that I should be sorry to lessen them by appearing to give the preference to any other dog. At the same time I may be allowed to add, that I have seen such courage, perseverance, and fidelity in the Newfoundland dog, and am acquainted with so many well-authenticated facts of his more than ordinary sense and utility, that I think him entitled to be considered as little inferior to the Irish wolf-dog.

When we reflect on the docility of the Newfoundland dog, his affectionate disposition, his aptitude in receiving instruction, and his instantaneous sense of impending danger, we shall no longer wonder at his being called the friend of his master, whom he is at all times ready to defend at the risk of his own life. How noble is his appearance, and at the same time how serene is his countenance!

> "Sa fierté, sa beauté, sa jeunesse agréable
> Le fit cherir de vous, et il est redoutable
> A vos fiers ennemis par sa courage."

Page 135 No animal, perhaps, can show more real courage than this dog. His perseverance in what he undertakes is so great, that he never relinquishes an attempt which has been enjoined him as long as there is a chance of success. I allude more particularly to storms at sea and consequent shipwreck, when his services, his courage, and indefatigable exertions, have been truly wonderful. Numerous persons have been saved from a watery grave by these dogs, and ropes have been conveyed by them from a sinking ship to the shore amidst foaming billows, by which means whole crews have been saved from destruction. Their feet are particularly well adapted to enable them to swim, being webbed very much like those of a duck, and they are at all times ready to plunge into the water to save a human being from drowning. Some dogs delight in following a fox, others in hunting the hare, or killing vermin. The delight of the Newfoundland dog appears to be in the preservation of the lives of the human race. A story is related on good authority of one of these dogs being in the habit, when he saw persons swimming in the Seine at Paris, of seizing them and bringing them to the shore. In the immediate neighbourhood of Windsor a servant was saved from drowning by a Newfoundland dog, who seized him by the collar of his coat when he was almost exhausted, and brought him to the banks, where some of the family were assembled watching with great anxiety the exertions of the noble animal.

Page 136 Those who were much at Windsor, not many years since, must have seen a fine Newfoundland dog, called Baby, reposing occasionally in front of the White Hart Hotel. Baby was a general favourite, and he deserved to be so; for he was mild in his disposition, brave as a lion, and very sensible. When he was thirsty, and could not procure water at the pump in the yard, he has frequently been seen to go to the stable, fetch an empty bucket, and stand with it in his mouth at the pump till some one came for water. He then, by wagging his tail and expressive looks, made his want known, and had his bucket filled. Exposed as Baby was to the attacks of all sorts of curs, as he slumbered in the sun in front of the hotel, he seemed to think that a pat with his powerful paw was quite sufficient punishment for them, but he never tamely submitted to insult from a dog approaching his own size, and his courage was only equalled by his gentleness.

The following anecdote, which is well authenticated, shows the sagacity as well as the kindliness of disposition of these dogs. In the city of Worcester, one of the principal streets leads by a gentle declivity to the river Severn. One day a child, in crossing the street, fell down in the middle of it, and a horse and cart, which were descending the hill, would have passed over it, had not a Newfoundland dog rushed to the rescue of the child, caught it up in his mouth, and conveyed it in safety to the foot pavement.

Page 137 My kind friend, Mr. T— —, took a Newfoundland dog and a small spaniel into a boat with him on the river Thames, and when he got into the middle of the river, he turned them into the water. They swam different ways, but the spaniel got into the current, and after struggling some time was in danger of being drowned. As soon as the Newfoundland dog perceived the predicament of his companion, he swam to his assistance, and brought him safe to the shore.

A vessel went down in a gale of wind near Liverpool, and every one on board perishes. A Newfoundland dog was seen swimming about the place where the vessel was lost for some time, and at last came on shore very much exhausted. For three days he swam off to the same spot, and was evidently trying to find his lost master, so strong was his affection.

I have always been pleased with that charming remark of Sir Edwin Landseer, that the Newfoundland dog was a "distinguished Member of the Humane Society." How delightfully has that distinguished artist portrayed the character of dogs in his pictures! and what justice has he done to their noble qualities! We see in them honesty, fidelity, courage, and sense—no exaggeration—no flattery. He makes us feel that his dogs will love us without selfishness, and defend us at the risk of their own lives—that though friends may forsake us, they never will—and that in misfortune, poverty, and death, their affection will be unchanged, Page 138 and their gratitude unceasing. But to return to the Newfoundland dog, and we shall again find him acting his part as a Member of the Humane Society.

A gentleman bathing in the sea at Portsmouth, was in the greatest danger of being drowned. Assistance was loudly called for, but no boat was ready, and though many persons were looking on, no one could be found to go to his help. In this predicament, a Newfoundland dog rushed into the sea and conveyed the gentleman in safety to land. He afterwards purchased the dog for a large sum, treated him as long as he lived with gratitude and kindness, and had the following words worked on his table-cloths and napkins—"*Virum extuli mari.*"

A person, in crossing a plank at a mill, fell into the stream at night, and was saved by his Newfoundland dog, and who afterwards recovered his hat, which had fallen from his head, and was floating down the stream.

There can be no doubt but that dogs calculate, and almost reason. A dog who had been in the habit of stealing from a kitchen, which had two doors opening into it, would never do so if one of them was shut, as he was afraid of being caught. If both the doors were open, his chance of escape was greater, and he therefore seized what he could. This sort of calculation, if I may call it is so, was shown by a Newfoundland bitch. She had suckled two whelps until they were able to take care of themselves. They were, Page 139 however, constantly following and disturbing her in order to be suckled, when she had little or no milk to give them. She was confined in a shed, which was separated from another by a wooden partition some feet high. Into this shed she conveyed her puppies, and left them there while she returned to the other to enjoy a night's rest unmolested. This shows that the animal was capable of reflecting to a degree beyond what would have been the result of mere instinct.

The late Rev. James Simpson, of the Potterrow congregation, Edinburgh, had a large dog of the Newfoundland breed. At that time he lived at Libberton, a distance of two miles from Edinburgh, in a house to which was attached a garden. One Sacrament Sunday the servant, who was left at home in charge of the house, thought it a good opportunity to entertain her friends, as her master and mistress were not likely to return home till after the evening's service, about nine o'clock. During the day the dog accompanied them through the garden, and indeed wherever they went, in the most attentive manner, and seemed well pleased. In the evening, when the time arrived that the party meant to separate, they proceeded to do so; but the dog, the instant they went to the door, interposed, and placing himself before it, would not allow one of them to touch the handle. On their persisting and attempting to use force he became furious, and in a menacing manner drove them back into the kitchen, where he kept them until the arrival of Mr. and Mrs. Simpson, Page 140 who were surprised to find the party at so late an hour, and more so to see the dog standing sentinel over them. Being thus detected, the servant acknowledged the whole circumstance, when her friends were allowed to depart, after being admonished by the worthy divine in regard to the proper use of the Sabbath. They could not but consider the dog as an instrument in the hand of Providence to point out the impropriety of spending this holy day in feasting rather than in the duties of religion.

After the above circumstance, it became necessary for Mr. Simpson, on account of his children's education, to leave his country residence, when he took a house in Edinburgh in a common stair. Speaking of this, one day, to a friend who had visited him, he concluded that he would be obliged to part with his dog, as he was too large an animal to be kept in such a house. The animal was present, and heard him say so, and must have understood what he meant, as he disappeared that evening, and was never afterwards heard of. These circumstances have been related to me by an elder of Mr. Simpson's congregation, who had them from himself.

I am indebted to the late amiable Lord Stowell for the following anecdote, which has since been verified by Mr. Henry Wix, brother of the archdeacon: —

A Newfoundland dog belonging to Archdeacon Wix, which had never quitted the island, was brought Page 141 over to London by him in January 1834, and when he and his family landed at Blackwall the dog was left on board the vessel. A few days afterwards the Archdeacon went from the Borough side of the Thames in a boat to the vessel, which was then in St. Katherine's Docks, to see about his luggage, but did not intend at that time to take the dog from the ship; however, on his leaving the vessel the dog succeeded in extricating himself from his confinement, jumped overboard, and swam after the boat across the Thames, followed his master into a counting-house on Gun-shot Wharf, Tooley Street, and then over London Bridge and through the City to St. Bartholomew's Hospital. The dog was shut within the square whilst the Archdeacon went into his father's house, and he then followed him on his way to Russell Square, but strayed somewhere in Holborn; and as several gentlemen had stopped to admire him in the street, saying he was worth a great deal of money, the Archdeacon concluded that some dog-stealer had enticed him away. He however wrote to the captain of the vessel to mention his loss, and made inquiries on the following morning at St. Bartholomew's Hospital, when he learnt that the dog had come to the gates late in the evening, and howled most piteously for admission, but was driven away. Two days afterwards the captain of the vessel waited on the Archdeacon with the dog, who had not only found his way back to the water's edge, on the Borough side, but, what is more surprising, swam across the Thames, Page 142 where no scent could have directed him, and found out the vessel in St. Katherine's Docks.

This sagacious and affectionate creature had, previous to his leaving Newfoundland, saved his master's life by directing his way home when lost in a snow-storm many miles from any shelter.

The dog was presented to the Archdeacon's uncle, Thomas Poynder, Esq., Clapham Common, in whose possession it continued until its death.

Every particular has been faithfully given of this extraordinary occurrence. Here we see a dog brought for the first time from Newfoundland, and who can scarcely be said to have put his feet on ground in England, not only finding his way through a crowded city to the banks of the river, but also finding the ship he wanted in that river, and in which he evidently thought he should discover his lost master. It is an instance of sense of so peculiar a kind that it is difficult to define it, or the faculty which enables animals to find their way to a place over ground which they had not previously traversed.

A gentleman of Suffolk, on an excursion with his friend, was attended by a Newfoundland dog, which soon became the subject of conversation. The master, after a warm eulogium upon the perfections of his canine favourite, assured his companion that he would, upon receiving the order, return and fetch any article Page 143 he should leave behind, from any distance. To confirm this assertion, a marked shilling was put under a large square stone by the side of the road, being first shown to the dog. The gentlemen then rode for three miles, when the dog received his signal from the master to return for the shilling he had seen put under the stone. The dog turned back; the gentlemen rode on, and reached home; but to their surprise and disappointment the hitherto faithful messenger did not return during the day. It afterwards appeared that he had gone to the place where the shilling was deposited, but the stone being too large for his strength to remove, he had stayed howling at the place till two horsemen riding by, and attracted by his seeming distress, stopped to look at him, when one of them alighting, removed the stone, and seeing the shilling, put it into his pocket, not at the time conceiving it to be the object of the dog's search. The dog followed their horses for twenty miles, remained undisturbed in the room where they supped, followed the chambermaid into the bedchamber, and secreted himself under one of the beds. The possessor of the shilling hung his trousers upon a nail by the bed-side; but when the travellers were both asleep, the dog took them in his mouth, and leaping out of the window, which was left open on account of the sultry heat, reached the house of his master at four o'clock in the morning with the prize he had made free with, in the pocket of which were found a watch and money, that were returned Page 144 upon being advertised, when the whole mystery was mutually unravelled, to the admiration of all the parties. [I]

Many years ago, I saw a horse belonging to a quartermaster in the 1st Dragoon Guards, when the regiment was quartered at Ipswich, find a shilling, which was covered with sawdust, in the riding-school at the Cavalry Barracks at that place, and give it to his owner. I thought this a wonderful instance of sagacity as well as docility, but how very far does this fall short of the intellectual faculty of dogs! I do not intend to assert that they are endowed with mental powers equal to those which the human race possess, but to contend that there is not a faculty of the human mind of which some evident proofs of its existence may not be found in dogs. Thus we find them possessed of memory, imagination, the powers of imitation, curiosity, cunning, revenge, ingenuity, gratitude, devotion, or affection, and other qualities. They are able to communicate their wants, their pleasures, and their pains, their apprehensions of danger, and their prospects of future good, by modulating their voices accordingly, and by significant gestures. They perfectly comprehend our wishes, and live with us as friends and companions.

When the fear of man and dread of him were inflicted as a curse on the animal creation, the dog-kind alone seems an exception, and their sagacity and fidelity to Page 145 the human race was an incalculable blessing bestowed upon them. These remarks are fully borne out in a very interesting article on the dog in the "Quarterly Review" of September, 1843.

A fine, handsome, and valuable black dog of the Newfoundland species, belonging to Mr. Floyd, solicitor, Holmfirth, committed suicide by drowning itself in the river which flows at the back of its owner's habitation. For some days previous the animal seemed less animated than usual, but on this particular occasion he was noticed to throw himself into the water and endeavour to sink by preserving perfect stillness of the legs and feet. Being dragged out of the stream, the dog was tied up for a time, but had no sooner been released than he again hastened to the water and again tried to sink, and was again got out. This occurred many times, until at length the animal with repeated efforts appeared to get exhausted, and by dint of keeping his head determinedly under water for a few minutes succeeded at last in obtaining his object, for when taken out this time he was indeed dead. The case is worth recording, as affording another proof of the general instinct and sagacity of the canine race.

Mr. Nicol, late of Pall Mall, told me he saw an old foxhound deliberately drown itself, and was ready to make oath of it.

Mrs. Kaye, residing opposite Windsor Park Wall, Datchet, had a beautiful Newfoundland dog. For the convenience of the family a boat was kept, that they Page 146 might at times cross the water without the inconvenience of going a considerable way round to Datchet Bridge. The dog was so delighted with the aquatic trips, that he very rarely permitted the boat to go without him. It happened that the coachman, who had been but little accustomed to the depths and shallows of the water, intending a forcible push with the punt pole, which was not long enough to reach the bottom, fell over the side of the boat in the deepest part of the water, and in the central part of the current, which accident was observed by a part of the family then at the front windows of the house; sudden and dreadful as the alarm was, they had the consolation of seeing the sagacious animal instantaneously follow his companion, when after diving, and making two or three abortive attempts, by laying hold of different parts of his apparel, which as repeatedly gave way or overpowered his exertions, he then, with the most determined and energetic fortitude, seized him by the arm, and brought him to the edge of the bank, where the domestics of the terrified family were ready to assist in extricating him from his perilous situation. [J]

I have mentioned that revenge had been shown by dogs, and the following is an instance of it. A gentleman was staying at Worthing, where his Newfoundland dog was teased and annoyed by a small cur, which snapped and barked at him. This he bore, without appearing to notice it, for some time; but at last the Page 147 Newfoundland dog seemed to lose his usual patience and forbearance, and he one day, in the presence of several spectators, took the cur up by his back, swam with it into the sea, held it under the water, and would probably have drowned it, had not a boat been put off and rescued it. There was another instance communicated to me. A fine Newfoundland dog had been constantly annoyed by a small spaniel. The former, seizing the opportunity when they were on a terrace under which a river flowed, took up the spaniel in his mouth, and dropped it over the parapet into the river.

Jukes, in his "Excursions in and about Newfoundland," says, "A thin, short-haired black dog, belonging to George Harvey, came off to us to-day; this animal was of a breed very different from what we understand by the term Newfoundland dog in England. He had a thin tapering snout, a long thin tail, and rather thin but powerful legs, with a lank body, the hair short and smooth. These are the most abundant dogs of the country, the long-haired curly dogs being comparatively rare. They are by no means handsome, but are generally more intelligent and useful than the others. This one caught his own fish; he sat on a projecting rock beneath a fish-lake or stage, where the fish are laid to dry, watching the water, which had a depth of six or eight feet, the bottom of which was white with fish-bones. On throwing a piece of codfish into the water, three or four heavy, clumsy-looking fish, called in Newfoundland sculpins, with great heads and mouths, Page 148 and many spines about them, and generally about a foot long, would swim in to catch it. These he would '*set*' attentively, and the moment one turned his broadside to him, he darted down like a fish-hawk, and seldom came up without the fish in his mouth. As he caught them he carried them regularly to a place a few yards off, where he laid them down; and they told us that in the summer he would sometimes make a pile of fifty or sixty a-day just at that place. He never attempted to eat them, but seemed to be fishing purely for his own amusement. I watched him for about two hours, and when the fish did not come I observed he once or twice put his right foot in the water, and paddled it about. This foot was white, and Harvey said he did it to *toll* or entice the fish; but whether it was for that specific reason, or merely a motion of impatience, I could not exactly decide."

Extraordinary as the following anecdote may appear to some persons, it is strictly true, and strongly shows the sense, and I am almost inclined to add, reason of the Newfoundland dog.

A friend of mine, while shooting wild fowl with his brother, was attended by a sagacious dog of this breed. In getting near some reeds by the side of a river, they threw down their hats, and crept to the edge of the water, when they fired at some birds. They soon afterwards sent the dog to bring their hats, one of which was smaller than the other. After several attempts to bring them both together in his mouth, the dog at last Page 149 placed the smaller hat in the larger one, pressed it down with his foot, and thus was able to bring them both at the same time.

A gentleman residing in Fifeshire, and not far from the city of St. Andrews, was in possession of a very fine Newfoundland dog, which was remarkable alike for its tractability and its trustworthiness. At two other points, each distant about a mile, and at the same distance from this gentleman's mansion, there were two dogs of great power, but of less tractable breeds than the Newfoundland one. One of these was a large mastiff, kept as a watch-dog by a farmer, and the other a stanch bull-dog, that kept guard over the parish mill. As each of these three was lord-ascendant of all animals at his master's residence, they all had a good deal of aristocratic pride and pugnacity, so that two of them seldom met without attempting to settle their respective dignities by a wager of battle.

The Newfoundland dog was of some service in the domestic arrangements, besides his guardianship of the house; for every forenoon he was sent to the baker's shop in the village, about half-a-mile distant, with a towel containing money in the corner, and he returned with the value of the money in bread. There were many useless and not over-civil curs in the village, as there are in too many villages throughout the country; but generally the haughty Newfoundland treated this ignoble race in that contemptuous style in which great Page 150 dogs are wont to treat little ones. When the dog returned from the baker's shop, he used to be regularly served with his dinner, and went peaceably on house-duty for the rest of the day.

One day, however, he returned with his coat dirtied and his ears scratched, having been subjected to a combined attack of the curs while he had charge of his towel and bread, and so could not defend himself. Instead of waiting for his dinner as usual, he laid down his charge somewhat sulkily, and marched off; and, upon looking after him, it was observed that he was crossing the intervening hollow in a straight line for the house of the farmer, or rather on an embassy to the farmer's mastiff. The farmer's people noticed this unusual visit, which they were induced to do from its being a meeting of peace between those who had habitually been belligerents. After some intercourse, of which no interpretation could be given, the two set off together in the direction of the mill; and having arrived there, they in brief space engaged the miller's bull-dog as an ally.

The straight road to the village where the indignity had been offered to the Newfoundland dog passed immediately in front of his master's house, but there was a more private and more circuitous road by the back of the mill. The three took this road, reached the village, scoured it in great wrath, putting to the tooth every cur they could get sight of; and having taken their revenge, and washed themselves in a ditch, they Page 151 returned, each dog to the abode of his master; and, when any two of them happened to meet afterwards, they displayed the same pugnacity as they had done previous to this joint expedition.

There is a well-authenticated anecdote of two dogs at Donaghadee, in which the instinctive daring of the one by the other caused a friendship, and, as it should seem, a kind of lamentation for the dead, after one of them had paid the debt of nature. This happened while the Government harbour or pier for the packets at Donaghadee was in the course of building, and it took place in the sight of several witnesses. The one dog in this case was also a Newfoundland, and the other was a mastiff. They were both powerful dogs; and though each was good-natured when alone, they were very much in the habit of fighting when they met. One day they had a fierce and prolonged battle on the pier, from the point of which they both fell into the sea; and as the pier was long and steep, they had no means of escape but by swimming a considerable distance. Throwing water upon fighting dogs is an approved means of putting an end to their hostilities; and it is natural to suppose that two combatants of the same species tumbling themselves into the sea would have the same effect. It had; and each began to make for the land as best he could. The Newfoundland being an excellent swimmer, very speedily gained the pier, on which he stood shaking himself; but at the same time watching the motions of his late antagonist, which, Page 152 being no swimmer, was struggling exhausted in the water, and just about to sink. In dashed the Newfoundland dog, took the other gently by the collar, kept his head above water, and brought him safely on shore. There was a peculiar kind of recognition between the two animals; they never fought again; they were always together: and when the Newfoundland dog had been accidentally killed by the passage of a stone waggon on the railway over him, the other languished and evidently lamented for a long time.

A gentleman had a pointer and Newfoundland dog, which were great friends. The former broke his leg, and was confined to a kennel. During that time the Newfoundland never failed bringing bones and other food to the pointer, and would sit for hours together by the side of his suffering friend.

During a period of very hot weather, the Mayor of Plymouth gave orders that all dogs found wandering in the public streets should be secured by the police, and removed to the prison-yard. Among them was a Newfoundland dog belonging to a shipowner of the port, who, with several others, was tied up in the yard. The Newfoundland soon gnawed the rope which confined him, and then hearing the cries of his companions to be released, he set to work to gnaw the ropes which confined them, and had succeeded in three or four instances, when he was interrupted by the entrance of the jailor.

A nearly similar case has frequently occurred in Page 153 the Cumberland Gardens, Windsor Great Park. Two dogs of the Newfoundland breed were confined in kennels at that place. When one of them was let loose, he has been frequently seen to set his companion free.

A boatman once plunged into the water to swim with another man for a wager. His Newfoundland dog, mistaking the purpose, and supposing that his master was in danger, plunged after him, and dragged him to the shore by his hair, to the great diversion of the spectators.

Mr. Peter Macarthur informs me, that in the year 1821, when opposite to Falmouth, he was at breakfast with a gentleman, when a large Newfoundland dog, all dripping with water, entered the room, and laid a newspaper on the table. The gentleman (who was one of the Society of Friends) informed the party, that this dog swam regularly across the ferry every morning, and went to the post-office, and fetched the papers of the day.

Mr. Blaine, in his "Encyclopædia of Rural Sports," tells the following story:—A Newfoundland dog, of the small, smooth-haired variety, in coming to England from his native country, was washed overboard during a tempestuous night. As daylight appeared the gale ceased, when a sailor at the mast-head descried something far in the wake of the vessel, which, by the help of his glass, he was led to believe was the dog, which was so great a favourite with the crew that it was unanimously requested of the captain of the vessel to *lie to,* and wait for the chance of saving the poor brute. Page 154 The captain, who had probably lost some time already by the storm, peremptorily refused to listen to the humane proposal. Whether it was the kindly feeling of the sailors, or the superstitious dread that if the dog were suffered to perish nothing would afterwards prosper with them, we are not informed; but we do know that, as soon as a refusal was made, the steersman left the helm, roundly asserting that he for one would never lend a hand to steer away from either Christian or brute in distress. The feeling was immediately caught by the rest of the crew, and maintained so resolutely, that the captain was forced to accede to the general wish; and the poor dog eventually reached the ship in safety, after having been, as we were informed, and implicitly believe, some hours in a tempestuous sea.

Bewick mentions an instance which shows the extraordinary sagacity of these dogs.

In a severe storm, a ship was lost off Yarmouth, and no living creature escaped, except a Newfoundland dog, which swam to the shore with the captain's pocket-book in his mouth. Several of the bystanders attempted to take it from him, but he would not part with it. At length, selecting one person from the crowd, whose appearance probably pleased him, he leaped against his breast in a fawning manner, and delivered the book to his care.

After mentioning this anecdote it will not be displeasing to read Lord Grenville's lines on his faithful Page 155 Newfoundland, as they may now be seen at Dropmore, with the translation of them:—

TIPPO.

In Villa.

Tippo ego hic jaceo, lapidem ne sperne, viator,
Qui tali impositus stat super ossa cani.
Larga mî natura manu dedit omnia, nostrum
Quæcunque exornant nobilitantque genus:
Robur erat validum, formæ concinna venustas,
Ingenui mores, intemerata fides.
Nec pudet invisi nomen gessisse tyranni,
Si tam dissimili viximus ingenio.
Naufragus in nuda Tenbeiæ [K] ejectus arena,
Ploravi domino me superesse meo,
Quem mihi, luctanti frustra, frustraque juvanti,
Abreptum, oceani in gurgite mersit hyems.
Solus ego sospes, sed quas miser ille tabellas
Morte mihi in media credidit, ore ferens.
Dulci me hospitio Belgæ excepere coloni,
Ipsa etiam his olim gens aliena plagis;
Et mihi gratum erat in longa spatiarier [L] ora,

Et quanquam infido membra lavare mari;
Gratum erat æstivis puerorum adjungere turmis
Participem lusus me, comitemque viæ.
Verum ubi, de multis captanti frustula mensis,
Bruma aderat, seniique hora timenda mei,
Insperata adeo illuxit fortuna, novique
Perfugium et requiem cura dedit domini.
Exinde hos saltus, hæc inter florea rura,
Et vixi felix, et tumulum hunc habeo.

Page 156

TIPPO.

Translated by a young Lady, a near Relation of the Author.

Here, stranger, pause, nor view with scornful eyes
The stone which marks where faithful Tippo lies.
Freely kind Nature gave each liberal grace,
Which most ennobles and exalts our race,
Excelling strength and beauty joined in me,
Ingenuous worth, and firm fidelity.
Nor shame I to have borne a tyrant's name,
So far unlike to his my spotless fame.
Cast by a fatal storm on Tenby's coast,
Reckless of life, I wailed my master lost.
Whom long contending with the o'erwhelming wave
In vain with fruitless love I strove to save.
I, only I, alas! surviving bore,
His dying trust, his tablets, [M] to the shore.
Kind welcome from the Belgian race I found,
Who, once in times remote, to British ground
Strangers like me came from a foreign strand.
I loved at large along the extended sand
To roam, and oft beneath the swelling wave,
Tho' known so fatal once, my limbs to lave;
Or join the children in their summer play,
First in their sports, companion of their way.

Thus while from many a hand a meal I sought,
Winter and age had certain misery brought;
But Fortune smiled, a safe and blest abode
A new-found master's generous love bestowed,
And midst these shades, where smiling flow'rets bloom,
Gave me a happy life and honoured tomb.

Dr. Abell, in one of his lectures on phrenology, related a very striking anecdote of a Newfoundland Page 157 dog at Cork. This dog was of a noble and generous disposition, and when he left his master's house was often assailed by a number of little noisy dogs in the street. He usually passed them with apparent unconcern, as if they were beneath his notice. One little cur, however, was particularly troublesome, and at length carried his petulance so far as to bite the Newfoundland dog in the back of his foot. This was too much to be patiently endured. He instantly turned round, ran after the offender, and seized him by the skin of his back. In this way he carried him in his mouth to the quay, and holding him some time over the water, at length dropped him into it. He did not seem, however, to wish to punish the culprit too much, for he waited a little while the poor animal, who was unused to that element, was not only well ducked, but near sinking, when he plunged in himself, and brought the other safe to land.

An officer, late in the 15th Hussars, informed me that he had witnessed a similar occurrence at St. Petersburg. These certainly are instances of a noble and generous disposition, as well as of great forbearance in not resenting an injury.

I may add the following instance of sagacity from the same quarter.

A vessel was driven by a storm on the beach of Lydd, in Kent. The surf was rolling furiously. Eight men were calling for help, but not a boat could be got Page 158 off to their assistance. At length a gentleman came on the beach, accompanied by his Newfoundland dog. He directed the attention of the noble animal to the vessel, and put a short stick into his mouth. The intelligent and courageous dog at once understood his meaning, and sprung into the sea, fighting his way through the foaming waves. He could not, however, get close enough to the vessel to deliver that with which he was charged, but the crew joyfully made fast a rope to another piece of wood, and threw it towards him. The sagacious dog saw the whole business in an instant; he dropped his own piece, and immediately seized that which had been cast to him; and then, with a degree of strength and determination almost incredible, he dragged it through the surge and delivered it to his master. By this means a line of communication was formed, and every man on board saved.

The keeper of a ferry on the banks of the Severn had a sagacious Newfoundland dog. If a dog was left behind by his owner in crossing, and was afraid of taking to the water, the Newfoundland dog has been frequently known to take the yelping animal in his mouth and convey it into the river. A person while rowing a boat, pushed his Newfoundland dog into the stream. The animal followed the boat for some time, till, probably finding himself fatigued, he endeavoured to get into it by placing his feet on the side. His owner repeatedly pushed the dog away, and in one of Page 159 his efforts to do so he overbalanced himself and fell into the river, and would probably have been drowned, had not the noble and generous animal immediately seized and held him above water till assistance arrived from the shore.

About twelve years ago a fine dog of a cross-breed, between a Newfoundland and a pointer, had been left by the captain of a vessel in the care of Mr. Park, of the White Hart Inn, Greenock. A friend of his, a gentleman from Argyllshire, took a fancy to this dog; and, when returning home, requested the loan of him for some time from Mr. Park, which he granted. This gentleman had some time before married a lady much to the dissatisfaction of his friends, who, in consequence, treated her with some degree of coldness and neglect. While he remained at home, the dog constantly attended him, and paid no apparent attention to the lady, who, on her part, never evinced any particular partiality for the dog. One time, however, the gentleman was called from home on business, and was to be absent several days. He wished to take the dog with him; but no entreaties could induce him to follow. The animal was then tied up to prevent his leaving the house in his absence; but he became quite furious till he was released, when he flew into the house and found his mistress, and would not leave her. He watched at the door of whatever room she was in, and would allow no one to approach without her special permission. When the gentleman returned home, the Page 160 dog seemed to take no more notice of the lady, but returned quietly to his former lodging in the stable. The whole circumstance caused considerable surprise; and the gentleman, wishing to try if the dog would again act in the same manner, left home for a day or two, when the animal actually resumed the faithful guardianship of his mistress as before; and this he continued to do whenever his master was absent, all the time he remained in his possession, which was two years.

The following anecdotes of an astonishing dog called Dandie are related by Captain Brown: —

"Mr. M'Intyre, patent-mangle manufacturer, Regent Bridge, Edinburgh, has a dog of the Newfoundland breed, crossed with some other, named Dandie, whose sagacious qualifications are truly astonishing and almost incredible. As the animal continues daily to give the most striking proofs of his powers, he is well known in the neighbourhood, and any person may satisfy himself of the reality of those feats, many of which the writer has himself had the pleasure to witness.

"When Mr. M'Intyre is in company, how numerous soever it may be, if he but say to the dog, 'Dandie, bring me my hat,' he immediately picks out the hat from all the others, and puts it in his master's hand.

"Should every gentleman in company throw a penknife on the floor, the dog, when commanded, will select his master's knife from the heap, and bring it to him.

Page 161 "A pack of cards being scattered in the room, if his master have previously selected one of them, the dog will find it out and bring it to him.

"A comb was hid on the top of a mantel-piece in the room, and the dog required to bring it, which he almost immediately did, although in the search he found a number of articles, also belonging to his master, purposely strewed around, all which he passed over, and brought the identical comb which he was required to find, fully proving that he is not guided by the sense of smell, but that he perfectly understands whatever is spoken to him.

"One evening, some gentlemen being in company, one of them accidentally dropped a shilling on the floor, which, after the most careful search, could not be found. Mr. M'Intyre seeing his dog sitting in a corner, and looking as if quite unconscious of what was passing, said to him, 'Dandie, find us the shilling, and you shall have a biscuit.' The dog immediately jumped upon the table and laid down the shilling, which he had previously picked up without having been perceived.

"One time, having been left in a room in the house of Mrs. Thomas, High Street, he remained quiet for a considerable time; but as no one opened the door, he became impatient, and rang the bell; and when the servant opened the door, she was surprised to find the dog pulling the bell-rope. Since that period, which was the first time he was observed to do it, he pulls Page 162 the bell whenever he is desired; and what appears still more remarkable, if there is no bell-rope in the room, he will examine the table, and if he finds a hand-bell, he takes it in his mouth and rings it.

"Mr. M'Intyre having one evening supped with a friend, on his return home, as it was rather late, he found all the family in bed. He could not find his boot-jack in the place where it usually lay, nor could he find it anywhere in the room after the strictest search. He then said to his dog, 'Dandie, I cannot find my bootjack; search for it.' The faithful animal, quite sensible of what had been said to him, scratched at the room-door, which his master opened. Dandie proceeded to a very distant part of the house, and soon returned, carrying in his mouth the bootjack, which Mr. M. now recollected to have left that morning under a sofa.

"A number of gentlemen, well acquainted with Dandie, are daily in the habit of giving him a penny, which he takes to a baker's shop and purchases bread for himself. One of these gentlemen, who lives in James's Square, when passing some time ago, was accosted by Dandie, in expectation of his usual present. Mr. T— — then said to him, 'I have not a penny with me to-day, but I have one at home.' Having returned to his house some time after, he heard a noise at the door, which was opened by the servant, when in sprang Dandie to receive his penny. In a frolic Mr. T— — gave him a bad one, which he, as usual, carried to the baker, but was refused his bread, as the Page 163 money was bad. He immediately returned to Mr. T— —'s, knocked at the door, and when the servant opened it, laid the penny down at her feet, and walked off, seemingly with the greatest contempt.

"Although Dandie, in general, makes an immediate purchase of bread with the money which he receives, yet the following circumstance clearly demonstrates that he possesses more prudent foresight than many who are reckoned rational beings.

"One Sunday, when it was very unlikely that he could have received a present of money, Dandie was observed to bring home a loaf. Mr. M'Intyre being somewhat surprised at this, desired the servant to search the room to see if any money could be found. While she was engaged in this task, the dog seemed quite unconcerned till she approached the bed, when he ran to her, and gently drew her back from it. Mr. M. then secured the dog, which kept struggling and growling while the servant went under the bed, where she found 7½*d.* under a bit of cloth; but from that time he never could endure the girl, and was frequently observed to hide his money in a corner of a saw-pit, under the dust.

"When Mr. M. has company, if he desire the dog to see any one of the gentlemen home, it will walk with him till he reach his home, and then return to his master, how great soever the distance may be.

"A brother of Mr. M.'s and another gentleman went one day to Newhaven, and took Dandie along Page 164 with them. After having bathed, they entered a garden in the town; and having taken some refreshment in one of the arbours, they took a walk around the garden, the gentleman leaving his hat and gloves in the place. In the meantime some strangers came into the garden, and went into the arbour which the others had left. Dandie immediately, without being ordered, ran to the place and brought off the hat and gloves, which he presented to the owner. One of the gloves, however, had been left; but it was no sooner mentioned to the dog than he rushed to the place, jumped again into the midst of the astonished company, and brought off the glove in triumph.

"A gentleman living with Mr. M'Intyre, going out to supper one evening, locked the garden-gate behind him, and laid the key on the top of the wall, which is about seven feet high. When he returned, expecting to let himself in the same way, to his great surprise the key could not be found, and he was obliged to go round to the front door, which was a considerable distance about. The next morning strict search was made for the key, but still no trace of it could be discovered. At last, perceiving that the dog followed him wherever he went, he said to him, 'Dandie, you have the key—go, fetch it.' Dandie immediately went into the garden and scratched away the earth from the root of a cabbage, and produced the key, which he himself had undoubtedly hid in that place.

"If his master place him on a chair, and request Page 165 him to sing, he will instantly commence a howling, which he gives high or low as signs are made to him with the finger.

"About three years ago a mangle was sent by a cart from the warehouse, Regent Bridge, to Portobello, at which time the dog was not present. Afterwards, Mr. M. went to his own house, North Back of the Canongate, and took Dandie with him, to have the mangle delivered. When he had proceeded a little way the dog ran off, and he lost sight of him. He still walked forward; and in a little time he found the cart in which the mangle was, turned towards Edinburgh, with Dandie holding fast by the reins, and the carter in the greatest perplexity; the man stated that the dog had overtaken him, jumped on his cart, and examined the mangle, and then had seized the reins of the horse and turned him fairly round, and that he would not let go his hold, although he had beaten him with a stick. On Mr. M.'s arrival, however, the dog quietly allowed the carter to proceed to his place of destination."

The following is another instance of extraordinary sagacity. A Newfoundland dog, belonging to a grocer, had observed one of the porters of the house, and who was often in the shop, frequently take money from the till, and which the man was in the habit of concealing in the stable. The dog, having witnessed these thefts, became restless, pulling persons by the Page 166 skirts of their coats, and apparently wishing them to follow him. At length, an apprentice had occasion to go to the stable; the dog followed him, and having drawn his attention to the heap of rubbish under which the money was buried, began to scratch till he had brought the booty to view. The apprentice brought it to his master, who marked the money and restored it to the place where it had been hidden. Some of the marked money was soon afterwards found on the porter, who was taken before a magistrate, and convicted of the theft.

A Newfoundland dog, which was frequently to be seen in a tavern in the High Street of Glasgow, lay generally at the door. When any person came to the house, he trotted before them into an apartment, rang the bell, and then resumed his station at the door.

The great utility and sagacity of the Newfoundland dog, in cases of drowning, were shown in the following instance. Eleven sailors, a woman, and the waterman, had reached a sloop of war in Hamoaze in a shore-boat. One of the sailors, stooping rather suddenly over the side of the boat to reach his hat, which had fallen into the sea, the boat capsized, and they were all plunged into the water. A Newfoundland dog, on the quarter-deck of the sloop, seeing the accident, instantly leaped amongst the unfortunate persons, and seizing one man by the collar of his coat, he supported his head above water until a boat had hastened to the spot and saved the lives of all but the waterman. After delivering his Page 167 burden in safety, the noble animal made a wide circuit round the ship in search of another person; but not finding one, he took up an oar in his mouth which was floating away, and brought it to the side of the ship.

A sailor, attended by a Newfoundland dog, became so intoxicated, that he fell on the pavement in Piccadilly, and was unable to rise, and soon fell asleep. The faithful dog took a position at his master's head, and resisted every attempt made to remove him. The man, having at last slept off the fumes of his intoxicating libations, awoke, and being told of the care his dog had taken of him, exclaimed, "This is not the first time he has kept watch over me."

On Thursday evening, January 28, 1858, as the play of "Jessie Vere" was being performed at Woolwich Theatre, and when a scene in the third act had been reached, in which a "terrific struggle" for the possession of a child takes place between the fond mother and two "hired ruffians," a large Newfoundland dog, which had by some means gained admittance with its owner into the pit, leaped over the heads of the musicians in the orchestra, and flew to the rescue, seizing one of the assassins, and almost dragging him to the ground. It was with difficulty removed, and dragged off the stage. The dog, which is the property of the chief engineer of Her Majesty's ship Buffalo, has been habitually accustomed to the society of children, for whom he has on many occasions evinced strong proofs of affection.

Page 168 Mr. Bewick, in his history of Quadrupeds, mentions some instances of the sagacity and intellect of Newfoundland dogs; and it may not be uninteresting to the admirers of that celebrated wood-engraver to be informed, on the authority of his daughters, that the group on the bridge in his print of the Newfoundland dog represents Mr. Preston, a Printer of Newcastle, Mr. Vint, of Whittingham, Mr. Bell, House Steward, and Mr. Bewick. Their initials, P. V. B. and B., are introduced in the woodcut. The dog was drawn at Eslington, the seat of Mr. Liddell, the eldest son of Lord Ravensworth. [N]

In Newfoundland, this dog is invaluable, and answers the purpose of a horse. He is docile, capable of strong attachment, and is easy to please in the quality of his food, as he will live on scraps of boiled fish, either salted or fresh, and on boiled potatoes and cabbage. The natural colour of this dog is black, with the exception of a very few white spots. Their sagacity is sometimes so extraordinary, as on many occasions to show that they only want the faculty of speech to make themselves fully understood.

The Rev. L. Anspach, in his history of the Island of Newfoundland, mentions some instances of this intelligence.

One of the Magistrates of Harbour-Grace, the late Mr. Garland, had an old dog, which was in the habit of carrying a lantern before his master at night, as Page 169 steadily as the most attentive servant could do; stopping short when his master made a stop, and proceeding when he saw him disposed to follow him. If his master was absent from home, on the lantern being fixed to his mouth, and the command given, "Go, fetch your master," he would immediately set off and proceed directly to the town, which lay at the distance of more than a mile from the place of his master's residence. He would then stop at the door of every house which he knew his master was in the habit of frequenting, and, laying down his lantern, would growl and strike the door, making all the noise in his power until it was opened. If his master was not there, he would proceed further until he had found him. If he accompanied him only once into a house, it was sufficient to induce him to take that house in his round.

The principal use of this animal in Newfoundland, in addition to his qualities as a good watch-dog and a faithful companion, is to assist in fetching from the woods the *lumber* intended either for repairing the fish stages, or for fuel; and this is done by dragging it on the snow or ice, or else on sledges, the dog being tackled to it.

These animals bark only when strongly provoked. They are not quarrelsome, but treat the smaller species with a great degree of patience and forbearance. They will defend their masters on seeing the least appearance of an attack on his person. The well-known partiality of these dogs for the water, in Page 170 which they appear as if in their proper element, diving and keeping their heads under the surface for a considerable time, seems to give them some connexion with the class of amphibious animals. At the same time, the several instances of their superior sagacity, and the essential services which they have been frequently known to render to humanity, give them a distinguished rank in the scale of the brute creation. I will mention another instance of this.

The Durham packet of Sunderland was, in 1815, wrecked near Clay, in Norfolk. A faithful dog was employed to use his efforts to carry the lead-line on shore from the vessel; but there being a very heavy sea, and a deep beach, it appeared that the drawback of the surf was too powerful for the animal to contend with. Mr. Parker, ship-builder, of Wells, and Mr. Jackson, jun., of Clay, who were on the spot, observing this, instantly rushed into the sea, which was running very high, and gallantly succeeded, though at a great risk, in catching hold of the dog, which was much exhausted, but which had all this time kept the line in his mouth. The line being thus obtained, a communication with the vessel was established; and a warp being passed from the ship to the shore, the lives of all on board, nine in number, including two children, were saved.

Some dogs are of an extremely jealous disposition; and the following extraordinary instance of it was communicated to me by Mr. Charles Davis, the well-known Page 171 and highly-respected huntsman of Her Majesty's stag-hounds, a man who has gained many friends, and perhaps never lost one, by his well-regulated conduct and sporting qualifications.

He informed me that a friend of his had a fine Newfoundland dog, which was a great favourite with the family. While this dog was confined in the yard, a pet lamb was given to one of the children, which the former soon discovered to be sharing a great portion of those caresses which he had been in the habit of receiving. This circumstance produced so great an effect on the poor animal, that he refused to eat, and fretted till he became extremely unwell. Thinking that exercise might be of use to him, he was let loose. No sooner was this done, than the dog watched his opportunity, and seized the lamb in his mouth. He was seen conveying it down a lane, about a quarter of a mile from his master's house, at the bottom of which the river Thames flowed. On arriving at it, he held the lamb under water till it was drowned, and thus effectually got rid of his rival. On examining the lamb, it did not appear to have been bitten, or otherwise injured; and it might almost be supposed that the dog had chosen the easiest death in removing the object of his dislike.

The sense of these animals is, indeed, perfectly wonderful. A lieutenant in the navy informed me, that while his ship was under sail in the Mediterranean, a favourite canary bird escaped from its cage, and flew Page 172 into the sea. A Newfoundland dog on board witnessed the circumstance, immediately jumped into the sea, and swam to the bird, which he seized in his mouth, and then swam back with it to the ship. On arriving on board and opening the dog's mouth, it was found that the bird was perfectly uninjured, so tenderly had it been treated, as though the dog had been aware that the slightest pressure would have destroyed it.

Mr. Youatt, whose remarks on the usefulness and good qualities of the inferior animals, in his work on Humanity to Brutes, do him so much credit, gives the following anecdote as a proof of the reasoning power of a Newfoundland dog.

Wanting one day to go through a tall iron gate, from one part of his premises to another, he found a lame puppy lying just within it, so that he could not get in without rolling the poor animal over, and perhaps injuring it. Mr. Youatt stood for awhile hesitating what to do, and at length determined to go round through another gate. A fine Newfoundland dog, however, who had been waiting patiently for his wonted caresses, and perhaps wondering why his master did not get in as usual, looked accidentally down at his lame companion. He comprehended the whole business in a moment—put down his great paw, and as gently and quickly as possible rolled the invalid out of the way, and then drew himself back in order to leave room for the opening of the gate.

Page 173 We may be inclined to deny reasoning faculties to dogs; but if this was not reason, it may be difficult to define what else it could be.

Mr. Youatt also says, that his own experience furnishes him with an instance of the memory and gratitude of a Newfoundland dog, who was greatly attached to him. He says, as it became inconvenient to him to keep the dog, he gave him to one who he knew would treat him kindly. Four years passed, and he had not seen him; when one day, as he was walking towards Kingston, and had arrived at the brow of the hill where Jerry Abershaw's gibbet then stood, he met Carlo and his master. The dog recollected Mr. Youatt in a moment, and they made much of each other. His master, after a little chat, proceeded towards Wandsworth, and Carlo, as in duty bound, followed him. Mr. Youatt had not, however, got half-way down the hill when the dog was again at his side, lowly but deeply growling, and every hair bristling. On looking about, he saw two ill-looking fellows making their way through the bushes, which occupied the angular space between Roehampton and Wandsworth roads. Their intention was scarcely questionable, and, indeed, a week or two before, he had narrowly escaped from two miscreants like them. "I can scarcely say," proceeds Mr. Youatt, "what I felt; for presently one of the scoundrels emerged from the bushes, not twenty yards from me; but he no sooner saw my companion, and heard his growling, the loudness and Page 174 depth of which were fearfully increasing, than he retreated, and I saw no more of him or of his associate. My gallant defender accompanied me to the direction-post at the bottom of the hill, and there, with many a mutual and honest greeting, we parted, and he bounded away to overtake his rightful owner. We never met again; but I need not say that I often thought of him with admiration and gratitude."

It is pleasing to record such instances of kindness in a brute. Here we see a recollection of, and gratitude for, previous good treatment, and that towards one whom the dog had not seen for four years. There is a sort of bewilderment in the human mind, when we come to analyse the feelings, affections, and peculiar instinctive faculties of dogs. A French writer (Mons. Blaze) has asserted, that the dog most undoubtedly has all the qualities of a man possessed of good feeling, and adds that man has not the fine qualities of the dog. We make a virtue of that gratitude which is nothing more than a duty incumbent upon us, while it is an inherent quality in the dog.

"Canis gratus est, et amicitiæ memor."

We repudiate ingratitude, and yet every one is more or less guilty of it. Indeed, where shall we find the man who is free from it? Take, however, the first dog you meet with, and the moment he has adopted you for his master, from that moment you are sure of his Page 175 gratitude and affection. He will love you without calculating what he shall gain by it—his greatest pleasure will be to be near you—and should you be reduced to beg your bread, no poverty will induce him to abandon you. Your friends may, and probably will, do so—the object of your love and attachment will not, perhaps, like to encounter poverty with you. Your wife, by some possibility (it is a rare case, however, if she has received kind treatment) may forget her vows, but your dog will never leave you—he will either die at your feet, or if he should survive you, will accompany you to the grave.

An intelligent correspondent, to whom I am indebted for some sensible remarks on the faculties of dogs, has remarked that large-headed dogs are generally possessed of superior faculties to others. This fact favours the phrenological opinion that size of brain is evidence of superior power. He has a dog possessing a remarkably large head, and few dogs can match him in intelligence. He is a cross with the Newfoundland breed, and besides his cleverness in the field as a retriever, he shows his sagacity at home in the performance of several useful feats. One consists in carrying messages. If a neighbour is to be communicated with, the dog is always ready to be the bearer of a letter. He will take orders to the workmen who reside at a short distance from the house, and will scratch impatiently at their door when so employed, although at other times, desirous of sharing Page 176 the warmth of their kitchen fire, he would wait patiently, and then entering with a seriousness befitting the imagined importance of his mission, would carefully deliver the note, never returning without having discharged his trust. His usefulness in recovering articles accidentally lost has often been proved. As he is not always allowed to be present at dinner, he will bring a hat, book, or anything he can find, and hold it in his mouth as a sort of apology for his intrusion. He seems pleased at being allowed to lead his master's horse to the stable.

Newfoundland dogs may readily be taught to rescue drowning persons. In France, this forms a part of their education, and they are now kept in readiness on the banks of the Seine, where they form a sort of Humane Society Corps. By throwing the stuffed figure of a man into a river, and requiring the dog to fetch it out, he is soon taught to do so when necessary, and thus he is able to rescue drowning persons. This hint might not be thrown away on our own excellent Humane Society.

Many dogs are called of the Newfoundland breed who have but small relationship with that sensible animal. The St. John's and Labrador dogs are also very different from each other. The former is strong in his limbs, rough-haired, small in the head, and carries his tail very high. The other, by far the best for every kind of shooting, is oftener black than of another colour, and scarcely bigger than a pointer. Page 177 He is made rather long in the head and nose, pretty deep in the chest, very fine in the legs, has short or smooth hair, does not carry his tail so much curled as the other, and is extremely quick and active in running, swimming, or fighting. The St. John's breed of these dogs is chiefly used on their native coast by fishermen. Their sense of smelling is scarcely to be credited. Their discrimination of scent, in following a wounded pheasant through a whole covert full of game, appears almost impossible.

The real Newfoundland dog may be broken into any kind of shooting, and, without additional instruction, is generally under such command, that he may be safely kept in, if required to be taken out with pointers. For finding wounded game of every description there is not his equal in the canine race, and he is a *sine quâ non* in the general pursuit of wildfowl. These dogs should be treated gently, and much encouraged when required to do anything, as their faults are easily checked. If used roughly, they are apt to turn sulky. They will also recollect and avenge an injury. A traveller on horseback, in passing through a small village in Cumberland, observed a Newfoundland dog reposing by the side of the road, and from mere wantonness gave him a blow with his whip. The animal made a violent rush at and pursued him a considerable distance. Having to proceed through the same place the next journey, which was about twelve months afterwards, and while in the act Page 178 of leading his horse, the dog, no doubt recollecting his former assailant, instantly seized him by the boot, and bit his leg. Some persons, however, coming up, rescued him from further injury.

A gamekeeper had a Newfoundland dog which he used as a retriever. Shooting in a wood one day, he killed a pheasant, which fell at some distance, and he sent his dog for it. When half way to the bird, he suddenly returned, refusing to go beyond the place at which he had first stopped. This being an unusual circumstance, the man endeavoured more and more to enforce his command; which being unable to effect, either by words or his whip, he at last, in a great passion, gave the dog a violent kick in the ribs, which laid it dead at his feet. He then proceeded to pick up the bird, and on returning from the spot, discovered a man concealed in the thicket. He immediately seized him, and upon examination, several snares were found on his person. This may be a useful hint to those who are apt to take violent measures with their dogs.

A gentleman who had a country house near London, discovered on arriving at it one day that he had brought away a key, which would be wanted by his family in town. Having an intelligent Newfoundland dog, which had been accustomed to carry things, he sent him back with it. While passing with the key, the animal was attacked by a butcher's dog, against which he made no resistance, but got away from him. After safely delivering the key, he returned to rejoin Page 179 his master, but stopped in the way at the butcher's shop, whose dog again sallied forth. The Newfoundland this time attacked him with a fury, which nothing but revenge could have inspired, nor did he quit the aggressor till he had killed him.

The following fact affords another proof of the extraordinary sagacity of these dogs.

A Newfoundland dog of the true breed was brought from that country, and given to a gentleman who resided near Thames Street, in London. As he had no means of keeping the animal, except in close confinement, he sent him to a friend in Scotland by a Berwick smack. When he arrived in Scotland he took the first opportunity of escaping, and though he certainly had never before travelled one yard of the road, he found his way back to his former residence on Fishstreet Hill; but in so exhausted a state, that he could only express his joy at seeing his master, and then died.

So wonderful is the sense of these dogs, that I have heard of three instances in which they have voluntarily guarded the bed-chamber doors of their mistresses, during the whole night, in the absence of their masters, although on no other occasion did they approach them.

The Romans appear to have had a dog, which seems to have been very similar in character to our Newfoundland. In the Museum at Naples there is an antique bronze, discovered amongst the ruins of Her Page 180 culaneum, which represents two large dogs dragging from the sea some apparently drowned persons.

The following interesting fact affords another instance of the sagacity and good feeling of the Newfoundland dog:—

In the year 1841, as a labourer, named Rake, in the parish of Botley, near Southampton, was at work in a gravel-pit, the top stratum gave way, and he was buried up to his neck by the great quantity of gravel which fell upon him. He was at the same time so much hurt, two of his ribs being broken, that he found it impossible to make any attempt to extricate himself from his perilous situation. Indeed, nothing could be more fearful than the prospect before him. No one was within hearing of his cries, nor was any one likely to come near the spot. He must almost inevitably have perished, had it not been for a Newfoundland dog belonging to his employer. This animal had been watching the man at his work for some days, as if he had been aware that his assistance would be required; for no particular attachment to each other had been exhibited on either side. As soon, however, as the accident occurred, the dog jumped into the pit, and commenced removing the gravel with his paws; and this he did in so vigorous and expeditious a manner, that the poor man was at length able to liberate himself, though with extreme difficulty. What an example of kindness, sensibility, and I may add reason, does this instance afford us!

Page 181 A gentleman in Ireland had a remarkably fine and intelligent Newfoundland dog, named Boatswain, whose acts were the constant theme of admiration. On one occasion, an aged lady who resided in the house, and the mother-in-law of the owner of the dog, was indisposed and confined to her bed. The old lady was tired of chickens and other productions of the farmyard, and a consultation was held in her room as to what could be procured to please her fancy for dinner. Various things were mentioned and declined, in the midst of which Boatswain, who was greatly attached to the old lady, entered her room with a fine young rabbit in his mouth, which he laid at the foot of the bed, wagging his tail with great exultation. It is not meant to infer that the dog knew anything of the difficulty of finding a dinner to the lady's taste, but seeing her distressed in mind and body, it is not improbable that he had brought his offering in the hopes of pleasing her.

On another occasion, his master found this dog early one summer's morning keeping watch over an unfortunate countryman, who was standing with his back to a wall in the rear of the premises, pale with terror. He was a simple, honest creature, living in the neighbourhood. Having to attend some fair or market, about four o'clock in the morning, he made a short cut through the grounds, which were under the protection of Boatswain, who drove the intruder to the wall, and kept him there, showing his teeth, and giving Page 182 a growl whenever he offered to stir from the spot. In this way he was kept a prisoner till the owner of the faithful animal released him.

There was a Newfoundland dog on board H. M. S. Bellona, which kept the deck during the battle of Copenhagen, running backward and forward with so brave an anger, that he became a greater favourite with the men than ever. When the ship was paid off, after the peace of Amiens, the sailors had a parting dinner on shore. Victor was placed in the chair, and fed with roast beef and plum-pudding, and the bill was made out in Victor's name. This anecdote is taken from Southey's "Omniana."

I am indebted to a kind correspondent for the following anecdotes: —

"A friend of mine, who in the time of the war commanded the Sea Fencibles, in the neighbourhood of Southend, possessed in those days a magnificent Newfoundland dog, named Venture. This noble creature my friend was accustomed to take with him in the pursuit of wild fowl. One cold evening, after having tolerable sport, the dog was suddenly missed; he had been last seen when in pursuit of a winged bird. As the ice was floating in the river, and the dog was true to his name, and would swim any distance for the recovery of wounded game, it was feared he must have fallen a victim to the hazards of the sport, and his owner returned home in consequence much dispirited. On his arrival at his house, Page 183 what was his extreme surprise, on entering the drawing-room, to find his wife accompanied by the dog, and a fine mallard lying on the table: the lady had, on her part, been overwhelmed with anxiety by the dog's having returned alone some time before, knowing the frequently perilous amusement in which her husband had embarked. The dog had straight on his return rushed to the drawing-room where the lady sat, and had laid the wild duck at her feet, having brought it safely in his mouth several miles.

"A gentleman once sent a coat to the tailor to be mended — it was left upon a counter in the shop. His dog had accompanied the servant to the tailor's. The animal watched his opportunity, pulled the coat down from the counter, and brought it home in triumph to his master.

"There is a tendency in the pride of man to deny the power of reasoning in animals, while it is the belief of some that reason is often a more sure guide to the brute beast, for the purposes designed by Providence, than that of their detractors. The fact is, I think, few persons who reflect deny the power, in a degree, to the less gifted of Nature's works. Certainly not some of the wisest of our race. Bishop Butler in his 'Analogy,' I think, assumes it; while the following beautiful inscription, designed for the epitaph of a favourite Newfoundland dog, was penned by no less a person than the late wise and venerable Earl of Eldon: from it his views on this subject may, Page 184 I fancy, be easily discerned. They are published in the life of him, written by Horace Twiss:—

'You who wander hither,
Pass not unheeded
The spot where poor Cæsar
Is deposited.
* * * *
To his rank among created beings
The power of reasoning is denied!
Cæsar manifested joy,
For days before his master
Arrived at Encombe;
Cæsar manifested grief
For days before his master left it.
What name shall be given
To that faculty,
Which thus made expectation
A source of joy,
Which thus made expectation
A source of grief?'"

Page 185

THE COLLEY, OR SHEPHERD'S DOG.

"My dog (the trustiest of his kind)
With gratitude inflames my mind:
I mark his true, his faithful way,
And in my service copy Tray."—Gay.

Who that has seen has not been delighted with the charming picture by Mr. Landseer of the shepherd's dog, resting his head on the coffin which contained the body of his dead master! Grief, fidelity, and affection are so strongly portrayed in the countenance of the poor dog, that they cannot be mistaken. We may Page 186 fancy him to have been the constant companion of the old shepherd through many a dreary day of rain, and frost, and snow on the neighbouring hills, gathering the scattered flock with persevering industry, and receiving the reward of his exertions in the approbation of his master. On returning to the humble cottage at night, he partakes of the "shepherd's scanty fare;" and then, coiled up before the flickering light of a few collected sticks, cold and shivering with wet, he awakes to greet his master at the first glimmering of morn, and is ready to renew his toils. Poor dog! what a lesson do you afford to those who are incapable of your gratitude, fidelity, and affection! and what justice has the charming artist done to these noble qualities! I trust he will receive this fanciful description of his dog as a little tribute paid to his talents, as well as to his good feeling.

The late Mr. Satterthwaite, grandfather of Thomas Rogerson, Esq., of Liverpool and Ballamillaghyn, Isle of Man, who died some years ago at Coulthouse, near Hawkshead, soon after his marriage, resided near the Low Wood Inn, on the borders of Windermere Lake. He left home early one morning, accompanied by his shepherd's dog, to look after some sheep on the mountains near Rydal, about four miles distant; and discovering two at the bottom of a precipice between two rocks he descended, with the view of extricating them; but when he got to the bottom, he could neither assist them nor get up himself, and there he was confined Page 187 until midnight. The faithful dog remained at the top of the precipice watching his master; but at nightfall he proceeded home, scratched the door, and was let in by his mistress, who expressed her surprise at the barking of the dog and non-arrival of her husband. She had no sooner sat down than the dog ran barking towards her, and then went to the door: but as she did not follow, the dog ran to her again, seized her apron, and endeavoured to pull her to the door; which circumstance caused her to suppose some accident had befallen her husband. She immediately called up the servant-man, and told him she was sure, from the strange conduct of the dog, that something must have happened to his master. She told the man to take a lantern and some ropes, and follow the dog, taking care to get assistance at Ambleside; which he did. No sooner had the man opened the door than the dog bounded out, leaped up at him, barked, and then ran forward, but quickly returned, leaped up again, barked, and then ran forward, as if to hasten the man's speed. The faithful dog led the man and his companions to the prison of his master. The ropes were instantly lowered, and Mr. Satterthwaite was providentially released from his perilous situation. The sheep also were recovered.

How well do I recollect the Ettrick Shepherd descanting on the sagacity and perseverance of his favourite sheep-dog! His name was Sirrah, and he Page 188 told me the following extraordinary anecdote of him, which I give in his own words:—

"About seven hundred lambs, which were once under my care at weaning time, broke up at midnight, and scampered off in three divisions across the hills, in spite of all that I and an assistant lad could do to keep them together. 'Sirrah, my man!' said I in great affliction, 'they are awa'.' The night was so dark that I could not see Sirrah, but the faithful animal heard my words—words such as of all others were sure to set him most on the alert; and without much ado he silently set off in search of the recreant flock. Meanwhile I and my companion did not fail to do all in our power to recover our lost charge. We spent the whole night in scouring the hills for miles around, but of neither the lambs nor Sirrah could we obtain the slightest trace. It was the most extraordinary circumstance that had occurred in my pastoral life. We had nothing for it (day having dawned), but to return to our master, and inform him that we had lost his whole flock of lambs, and knew not what had become of them. On our way home, however, we discovered a body of lambs at the bottom of a deep ravine, called the Flesh Cleuch, and the indefatigable Sirrah standing in front of them, looking all around for some relief, but still standing true to his charge. The sun was then up; and when we first came in view of them, we concluded that it was one of the divi Page 189 sions which Sirrah had been unable to manage until he came to that commanding situation. But what was our astonishment, when we discovered by degrees that not one lamb of the whole flock was wanting! How he had got all the divisions collected in the dark, is beyond my comprehension. The charge was left entirely to himself, from midnight until the rising of the sun; and if all the shepherds in the forest had been there to have assisted him, they could not have effected it with greater propriety. All that I can farther say is, that I never felt so grateful to any creature below the sun, as I did to my honest Sirrah that morning."

"I once sent you," says Mr. Hogg, some years later, in a letter to the Editor of "Blackwood's Edinburgh Magazine," "an account of a notable dog of my own, named Sirrah, which amused a number of your readers a great deal, and put their faith in my veracity somewhat to the test; but in this district, where the singular qualities of the animal were known, so far from any of the anecdotes being disputed, every shepherd values himself to this day on the possession of facts far outstripping any of those recorded by you formerly. With a few of these I shall conclude this paper. But, in the first place, I must give you some account of my own renowned Hector, which I promised long ago. He was the son and immediate successor of the faithful old Sirrah; and though not nearly so valuable a dog as his father, he was a far more interesting one. He had Page 190 three times more humour and whim about him; and though exceedingly docile, his bravest acts were mostly tinctured with a grain of stupidity, which showed his reasoning faculty to be laughably obtuse.

"I shall mention a striking instance of it. I was once at the farm of Shorthope on Ettrick Head, receiving some lambs that I had bought, and was going to take to market, with some more, the next day. Owing to some accidental delay, I did not get final delivery of the lambs till it was growing late; and being obliged to be at my own house that night, I was not a little dismayed lest I should scatter and lose my lambs if darkness overtook me. Darkness did overtake me by the time I got half-way, and no ordinary darkness for an August evening. The lambs having been weaned that day, and of the wild black-faced breed, became exceedingly unruly, and for a good while I lost hopes of mastering them. Hector managed the point, and we got them safe home; but both he and his master were alike sore forefoughten. It had become so dark that we were obliged to fold them with candles; and, after closing them safely up, I went home with my father and the rest to supper. When Hector's supper was set down, behold he was awanting! and as I knew we had him at the fold, which was within call of the house, I went out and called and whistled on him for a good while, but he did not make his appearance. I was distressed about this; for, having to take away the lambs next morning, I knew Page 191 I could not drive them a mile without my dog if it had been to save the whole drove.

"The next morning, as soon as it was day, I arose and inquired if Hector had come home? No; he had not been seen. I knew not what to do; but my father proposed that he would take out the lambs and herd them, and let them get some meat to fit them for the road, and that I should ride with all speed to Shorthope to see if my dog had gone back there. Accordingly we went together to the fold to turn out the lambs, and there was poor Hector, sitting trembling in the very middle of the fold-door, on the inside of the flake that closed it, with his eyes still steadfastly fixed on the lambs. He had been so hardly set with them after it grew dark, that he durst not for his life leave them, although hungry, fatigued, and cold, for the night had turned out a deluge of rain. He had never so much as lain down; for only the small spot that he sat on was dry, and there had he kept watch the whole night. Almost any other colley would have discerned that the lambs were safe enough in the fold, but honest Hector had not been able to see through this. He even refused to take my word for it; for he would not quit his watch, though he heard me calling both at night and morning.

"Another peculiarity of his was, that he had a mortal antipathy to the family-mouser, which was ingrained in his nature from his very puppyhood; yet so perfectly absurd was he, that no impertinence on her Page 192 side, and no baiting on, could ever induce him to lay his mouth on her, or injure her in the slightest degree. There was not a day and scarcely an hour passed over, that the family did not get some amusement with these two animals. Whenever he was within doors, his whole occupation was watching and *pointing* the cat from morning to night. When she flitted from one place to another, so did he in a moment; and then squatting down, he kept his *point* sedulously, till he was either called off or fell asleep.

"He was an exceedingly poor eater of meat, always had to be pressed to it, and often would not take it till we brought in the cat. The malicious looks that he cast at her from under his eyebrows on such occasions were exceedingly ludicrous, considering his utter disinclination to injure her. Whenever he saw her, he drew near his bicker and looked angry; but still he would not taste till she was brought to it, and then he cocked his tail, set up his birses, and began lapping furiously as if in utter desperation. His good nature, however, was so immovable, that he would never refuse her a share of what was placed before him; he even lapped close to the one side of the dish, and left her room,—but mercy! how he did ply!

"It will appear strange to you to hear a dog's reasoning faculty mentioned as I have done; but I declare I have hardly ever seen a shepherd's dog do anything without believing that I perceived his reasons for it. I Page 193 have often amused myself in calculating what his motives were for such and such things, and I generally found them very cogent ones. But Hector had a droll stupidity about him, and took up forms and rules of his own, for which I could never perceive any motive that was not even farther out of the way than the action itself. He had one uniform practice, and a very bad one it was; during the time of family worship, and just three or four seconds before the conclusion of the prayer, he started to his feet and ran barking round the apartment like a crazed beast. My father was so much amused with this, that he would never suffer me to correct him for it, and I scarcely ever saw the old man rise from the prayer without his endeavouring to suppress a smile at the extravagance of Hector. None of us ever could find out how he knew that the prayer was near done, for my father was not formal in his prayers; but certes he did know,—and of that we had nightly evidence. There never was anything for which I was so puzzled to discover a motive as this, but from accident I did discover it; and, however ludicrous it may appear, I am certain I was correct. It was much in character with many of Hector's feats, and rather, I think, the most *outré* of any principle he ever acted on. As I said, his great daily occupation was pointing the cat. Now, when he saw us kneel all down in a circle, with our faces couched on our paws, in the same posture with himself, it struck his absurd head that we were all engaged in pointing the cat. Page 194 He lay on tenters all the while, but the acuteness of his ear enabling him, through time, to ascertain the very moment when we would all spring to our feet, he thought to himself, 'I shall be first after her, for you all.'

"He inherited his dad's unfortunate ear for music, not perhaps in so extravagant a degree, but he ever took care to exhibit it on the most untimely and ill-judged occasions. Owing to some misunderstanding between the minister of the parish and the session-clerk, the precenting in church devolved on my father, who was the senior elder. Now, my father could have sung several of the old church-tunes middling well in his own family-circle; but it so happened that, when mounted in the desk, he never could command the starting notes of any but one (St. Paul's), which were always in undue readiness at the root of his tongue, to the exclusion of every other semibreve in the whole range of sacred melody. The minister gave out psalms four times in the course of every day's service; consequently the congregation were treated with St. Paul's in the morning at great length, twice in the course of the service, and then once again at the close. Nothing but St. Paul's. And it being itself a monotonous tune, nothing could exceed the monotony that prevailed in the primitive church of Ettrick. Out of pure sympathy for my father alone, I was compelled to take the precentorship in hand; and having plenty of tunes, for a good while I came on as well as could Page 195 be expected, as men say of their wives. But, unfortunately for me, Hector found out that I attended church every Sunday, and though I had him always closed up carefully at home, he rarely failed in making his appearance in church at some time of the day. Whenever I saw him a tremor came over my spirits, for I well knew what the issue would be. The moment that he heard my voice strike up the psalm 'with might and majesty,' then did he fall in with such overpowering vehemence, that he and I seldom got any to join in the music but our two selves. The shepherds hid their heads, and laid them down on the backs of their seats rowed in their plaids, and the lasses looked down to the ground and laughed till their faces grew red. I despised to *stick* the tune, and therefore was obliged to carry on in spite of the obstreperous accompaniment; but I was, time after time, so completely put out of all countenance with the brute, that I was obliged to give up my office in disgust, and leave the parish once more to their old friend, St. Paul.

"Hector was quite incapable of performing the same feats among sheep that his father did; but, as far as his judgment served him, he was a docile and obliging creature. He had one singular quality, of keeping true to the charge to which he was set. If we had been shearing, or sorting sheep in any way, when a division was turned out and Hector got the Page 196 word to attend to them, he would have done it pleasantly for a whole day without the least symptom of weariness. No noise or hurry about the fold, which brings every other dog from his business, had the least effect on Hector, save that it made him a little troublesome on his own charge, and set him a-running round and round them, turning them in at corners, from a sort of impatience to be employed as well as his baying neighbours at the fold. Whenever old Sirrah found himself hard set in commanding wild sheep on steep ground, where they are worst to manage, he never failed, without any hint to the purpose, to throw himself wide in below them, and lay their faces to the hill, by which means he got the command of them in a minute. I never could make Hector comprehend this advantage with all my art, although his father found it out entirely of himself. The former would turn or wear sheep no other way but on the hill above them; and, though very good at it, he gave both them and himself double the trouble and fatigue.

"It cannot be supposed that he could understand all that was passing in the little family circle, but he certainly comprehended a good part of it. In particular, it was very easy to discover that he rarely missed aught that was said about himself, the sheep, the cat, or of a hunt. When aught of that nature came to be discussed, Hector's attention and impatience soon be Page 197 came manifest. There was one winter evening I said to my mother that I was going to Bowerhope for a fortnight, for that I had more conveniency for writing with Alexander Laidlaw than at home; and I added, 'But I will not take Hector with me, for he is constantly quarrelling with the rest of the dogs, singing music, or breeding some uproar.' 'Na, na,' quoth she, 'leave Hector with me; I like aye best to have him at hame, poor fallow.'

"These were all the words that passed. The next morning the waters were in a great flood, and I did not go away till after breakfast; but when the time came for tying up Hector, he was a-wanting. 'The deil's in that beast,' said I,—'I will wager that he heard what we were saying yesternight, and has gone off for Bowerhope as soon as the door was opened this morning.'

"'If that should really be the case, I'll think the beast no canny,' said my mother.

"The Yarrow was so large as to be quite impassable, so that I had to walk up by St. Mary's Loch, and go across by the boat; and, on drawing near to Bowerhope, I soon perceived that matters had gone precisely as I suspected. Large as the Yarrow was, and it appeared impassable by any living creature, Hector had made his escape early in the morning, had swam the river, and was sitting, 'like a drookit hen,' on a knoll at the east end of the house, awaiting my arrival with great impatience. I had a great attach Page 198 ment to this animal, who, to a good deal of absurdity, joined all the amiable qualities of his species. He was rather of a small size, very rough and shagged, and not far from the colour of a fox.

"His son Lion was the very picture of his dad, had a good deal more sagacity, but also more selfishness. A history of the one, however, would only be an epitome of that of the other. Mr. William Nicholson [O] took a fine likeness of this latter one, which he still possesses. He could not get him to sit for his picture in such a position as he wanted, till he exhibited a singularly fine portrait of a small dog, on the opposite side of the room. Lion took it for a real animal, and, disliking its fierce and important look exceedingly, he immediately set up his ears and his shaggy birses, and, fixing a stern eye on the picture in manifest wrath, he would then sit for a whole day and point at it without budging or altering his position.

"It is a curious fact in the history of these animals, that the most useless of the breed have often the greatest degree of sagacity in trifling and useless matters. An exceedingly good sheep-dog attends to nothing else but that particular branch of business to which he is bred. His whole capacity is exerted and exhausted on it, and he is of little avail in miscellaneous matters; whereas, a very indifferent cur, bred Page 199 about the house, and accustomed to assist in every thing, will often put the more noble breed to disgrace in those paltry services. If one calls out, for instance, that the cows are in the corn, or the hens in the garden, the house-colley needs no other hint, but runs and turns them out. The shepherd's dog knows not what is astir; and, if he is called out in a hurry for such work, all that he will do is to break to the hill, and rear himself up on end to see if no sheep are running away. A bred sheep-dog, if coming hungry from the hills, and getting into a milk-house, would most likely think of nothing else than filling his belly with the cream. Not so his uninitiated brother; he is bred at home to far higher principles of honour. I have known such lie night and day among from ten to twenty pails full of milk, and never once break the cream of one of them with the tip of his tongue, nor would he suffer cat, rat, or any other creature to touch it. This latter sort, too, are far more acute at taking up what is said in a family.

"The anecdotes of these animals are all so much alike, that were I but to relate the thousandth part of those I have heard, they would often look very much like repetitions. I shall therefore, in this paper, only mention one or two of the most singular, which I know to be well authenticated.

"There was a shepherd lad near Langholm, whose name was Scott, who possessed a bitch famed over all the West Border for her singular tractability. He Page 200 could have sent her home with one sheep, two sheep, or any given number, from any of the neighbouring farms; and, in the lambing season, it was his uniform practice to send her home with the kebbed ewes just as he got them. I must let the town reader understand this. A kebbed ewe is one whose lamb dies. As soon as such is found, she is immediately brought home by the shepherd, and another lamb put to her; and Scott, on going his rounds on the hill, whenever he found a kebbed ewe, immediately gave her in charge to his bitch to take home, which saved him from coming back that way again and going over the same ground he had visited before. She always took them carefully home, and put them into a fold which was close by the house, keeping watch over them till she was seen by some one of the family; upon which she instantly decamped, and hastened back to her master, who sometimes sent her three times home in one morning with different charges. It was the custom of the farmer to watch her and take the sheep in charge from her: but this required a good deal of caution; for as soon as she perceived that she was seen, whether the sheep were put into the fold or not, she concluded her charge was at an end, and no flattery could induce her to stay and assist in folding them. There was a display of accuracy and attention in this that I cannot say I have ever seen equalled.

"The late Mr. Steel, flesher in Peebles, had a bitch that was fully equal to the one mentioned above, and Page 201 that, too, in the very same qualification. Her feats in taking sheep from the neighbouring farms into the Flesh-market at Peebles, form innumerable anecdotes in that vicinity. But there is one related of her, that manifests so much sagacity with natural affection, that I do not think the history of the animal creation furnishes such another.

"Mr. Steel had such implicit dependence on the attention of this animal to his orders, that, whenever he put a lot of sheep before her, he took a pride in leaving them to herself, and either remained to take a glass with the farmer of whom he had made the purchase, or took another road to look after bargains or other business. But one time he chanced to commit a drove to her charge at a place called Willenslee, without attending to her condition as he ought to have done. This farm is five miles from Peebles, over wild hills, and there is no regularly defined path to it. Whether Mr. Steel remained behind, or chose another road, I know not; but, on coming home late in the evening, he was astonished at hearing that his faithful animal had not made her appearance with the flock. He and his son, or servant, instantly prepared to set out by different paths in search of her; but, on their going out to the street, there was she coming with the drove, no one missing; and, marvellous to relate, she was carrying a young pup in her mouth! She had been taken in travail on those hills; and how the poor beast had contrived to manage the drove in her state of Page 202 suffering is beyond human calculation, for her road lay through sheep the whole way. Her master's heart smote him when he saw what she had suffered and effected: but she was nothing daunted; and having deposited her young one in a place of safety, she again set out full speed to the hills, and brought another and another, till she removed her whole litter one by one; but the last one was dead.

"The stories related of the dogs of sheep-stealers are fairly beyond all credibility. I cannot attach credit to some of them without believing the animals to have been devils incarnate, come to the earth for the destruction both of the souls and bodies of men. I cannot mention names, for the sake of families that still remain in the country; but there have been sundry men executed, who belonged to this district of the kingdom, for that heinous crime, in my own days; and others have absconded, just in time to save their necks. There was not one of these to whom I allude who did not acknowledge his dog to be the greatest aggressor. One young man in particular, who was, I believe, overtaken by justice for his first offence, stated, that after he had folded the sheep by moonlight, and selected his number from the flock of a former master, he took them out, and set away with them towards Edinburgh. But before he had got them quite off the farm, his conscience smote him, as he said (but more likely a dread of that which soon followed), and he quitted the sheep, letting them go again to the hill. He Page 203 called his dog off them, and mounting his pony, he rode away. At that time he said his dog was capering and playing around him, as if glad of having got free of a troublesome business; and he regarded him no more, till, after having rode about three miles, he thought again and again that he heard something coming up behind him. Halting, at length, to ascertain what it was, in a few minutes there comes his dog with the stolen animals, driving them at a furious rate to keep up with his master. The sheep were all smoking, and hanging out their tongues, and their guide was fully as warm as they. The young man was now exceedingly troubled, for the sheep having been brought so far from home, he dreaded there would be a pursuit, and he could not get them home again before day. Resolving, at all events, to keep his hands clear of them, he corrected his dog in great wrath, left the sheep once more, and taking colley with him, rode off a second time. He had not ridden above a mile, till he perceived that his assistant had again given him the slip; and suspecting for what purpose, he was terribly alarmed as well as chagrined; for daylight now approached, and he durst not make a noise calling on his dog, for fear of alarming the neighbourhood, in a place where they were both well known. He resolved therefore to abandon the animal to himself, and take a road across the country which he was sure the other did not know, and could not follow. He took that road, but being on horseback, he Page 204 could not get across the enclosed fields. He at length came to a gate, which he shut behind him, and went about half a mile farther, by a zigzag course, to a farmhouse, where both his sister and sweetheart lived; and at that place he remained until after breakfast time. The people of this house were all examined on the trial, and no one had either seen the sheep or heard them mentioned, save one man, who came up to the aggressor as he was standing at the stable-door, and told him that his dog had the sheep safe enough down at the Crooked Yett, and he needed not hurry himself. He answered, that the sheep were not his—they were young Mr. Thomson's, who had left them to his charge, and he was in search of a man to drive them, which made him come off his road.

"After this discovery, it was impossible for the poor fellow to get quit of them; so he went down and took possession of the stolen drove once more, carried them on, and disposed of them; and, finally, the transaction cost him his life. The dog, for the last four or five miles that he had brought the sheep, could have no other guide to the road his master had gone but the smell of his pony's feet. I appeal to every unprejudiced person if this was not as like one of the deil's tricks as an honest colley's.

"It is also well known that there was a notorious sheep-stealer in the county of Mid-Lothian, who, had it not been for the skins and the heads, would never have been condemned, as he could, with the greatest Page 205 ease, have proved an *alibi* every time suspicions were entertained against him. He always went by one road, calling on his acquaintances, and taking care to appear to everybody by whom he was known, while his dog went by another with the stolen sheep; and then, on the two felons meeting again, they had nothing more to do than turn the sheep into an associate's enclosure, in whose house the dog was well fed and entertained, and would have soon taken all the fat sheep on the Lothian edges to that house. This was likewise a female, a jet-black one, with a deep coat of soft hair, but smooth-headed, and very strong and handsome in her make. On the disappearance of her master she lay about the hills and places where he had frequented, but she never attempted to steal a drove by herself, nor the smallest thing for her own hand. She was kept some time by a relation of her master's, but never acting heartily in his service, soon came privately to an untimely end. Of this there is little doubt, although some spread the report that one evening, after uttering two or three loud howls, she instantly vanished! From such dogs as these, good Lord deliver us!"

The following is, perhaps, a still more extraordinary anecdote of the fidelity shown by a sheep-dog to its charge. It was communicated by Robert Murray, shepherd to Mr. Samuel Richmond, Path of Coudie, near Dunning, in Perthshire.

Murray had purchased for his master four score of sheep at the Falkirk Tryst, but having occasion to stop Page 206 another day, and confident in the faithfulness and sagacity of his colley, which was a female, he committed the drove to her care, with orders to drive them home,—a distance of about seventeen miles. The poor animal, when a few miles on the road, dropped two whelps, but, faithful to her charge, she drove the sheep on a mile or two further—then, allowing them to stop, returned for her pups, which she carried for about two miles in advance of the sheep. Leaving her pups, the colley again returned for the sheep, and drove them onwards a few miles. This she continued to do, alternately carrying her own young ones and taking charge of the flock, till she reached home. The manner of her acting on this occasion was afterwards gathered by the shepherd from various individuals, who had observed these extraordinary proceedings of the dumb animal on the road. However, when the colley reached her home, and delivered her charge, it was found that the two pups were dead. In this extremity, the instinct of the poor brute was, if possible, still more remarkable. She went to a rabbit-brae in the vicinity, and dug out of the earth two young rabbits, which she deposited on some straw in a barn, and continued to suckle for some time, until one of the farm servants unluckily let down a full sack upon them and smothered them.

The following anecdote is related by Captain Brown:—

A shepherd had driven a part of his flock to a neighbouring farm, leaving his dog to watch the re Page 207 mainder during that day and the next night, expecting to revisit them the following morning. Unfortunately, however, when at the fair, the shepherd forgot both his dog and his sheep, and did not return home till the morning of the third day. His first inquiry was, whether his dog had been seen? The answer was, No. "Then he must be dead," replied the shepherd in a tone of anguish, "for I know he was too faithful to desert his charge." He instantly repaired to the heath. The dog had sufficient strength remaining to crawl to his master's feet, and express his joy at his return, and almost immediately after expired.

Mr. Blaine relates the following circumstance: —I remember watching a shepherd boy in Scotland, who was sitting on the bank of a wide but shallow stream. A sheep had strayed to a considerable distance on the other side of the water; the boy, calling to his dog, ordered him to fetch that sheep back, but to do it gently, for she was heavy in lamb. I do not affect to say that the dog understood the reason for which he was commanded to perform this office in a more gentle manner than usual; but that he did understand he was to do it gently was very evident, for he immediately marched away through the water, came gently up to the side of the sheep, turned her towards the rest, and then they both walked quietly side by side to the flock. I was scarcely ever more pleased at a trifling incident in rural scenery than this.

Page 208 The sense and recollection of the sheep-dog were shown in the following instance: —

When I occupied a small farm in Surrey, I was in the habit of joining with a friend in the purchase of two hundred Cheviot sheep. The first year we had them, the shepherd who drove them from the North was asked by us how he had got on. "Why, very badly," said the man; "for I had a young dog, and he did not manage well in keeping the sheep from running up lanes and out-of-the-way places." The next year we had the same number of sheep brought up, and by the same man. In answer to our question about his journey, he informed us that he had got on very well, for his dog had recollected all the turnings of the road which the sheep had passed the previous year, and had kept them straight the whole of the way.

It has always appeared to me that the patriarchal flocks, the shepherds and their dogs, are seen to more advantage on the wild hills of Cumberland and Westmorland, than in any other situation. When I have wandered along the sides of some of the beautiful lakes of those counties, and have witnessed the effects of light and shade at different times of the day, on the water and distant hills and valleys, and seen the numerous sheep scattered over the latter, how delightful has been the prospect! During the early morning the bright beams of the sun did not produce too Page 209 much glare and heat, but served to give a charming glitter to the dew-drops as they besparkled the grass and flowers. The tracts of the sheep might be seen by the disappearance of the "gentle dew" from their path as they proceeded to their pasture, driven by the watchful colley. It was a scene of cheerfulness, which every lover of nature would admire.

In the evening the calmness of the lake was delightful. The light hovered over it, and the reflection of the trees in the transparent water beautified the scene. The beams of the setting sun glowed first over the valleys, and then illumined the tops of the hills; then gradually disappeared: but the grey tints of evening still had their beauty, and a diversity of them was preserved long after the greater effects of the setting sun had vanished. Deep shade was contrasted with former splendour, till at last the lovely moon appeared with her modest light, and formed a streak across the lake, which was occasionally broken as a ripple, raised by a breeze of the gentlest kind, passed over it.

While the sun still gleamed on the mountain's side the shepherd might be observed resting at its foot, while his patient dog ranged about collecting the flock, and bringing them towards his master.

Dear, lovely lake!—Never shall I forget your beauteous scenery. Seated in the cool of the evening under one of the noble trees on your shore, the only sounds I heard were the soft ripple of the water, and Page 210 the late warbling of the redbreast—Yes, I forget the humming beetle as it rapidly passed, and the owl calling to its mate in the distant wood. How peaceful were my feelings!—

"Happy the man whose tranquil mind
Sees Nature in her changes kind,
And pleased the whole surveys;
For him the morn benignly smiles,
And evening shades reward the toils
That measure out his days.

The varying year may shift the scene,
The sounding tempest lash the main,
And heaven's own thunder roll;
Calmly he views the bursting storm,
Tempests nor thunders can deform

The quiet of his soul." — C. B.

Nor is the scenery from the Lakes the only thing to be admired in this delightful country. Lanes may be traversed sheltered by the oak, the ash, and the hazel, and only those who have seen the Cumberland hazels can form an idea of the beauty of their silvery bark and luxuriant growth. From these lanes there are occasional openings, through which a placid lake or a distant range of hills may be seen. And what picturesque and rugged hills they are! Huge, projecting rocks and verdant lawns, and deep channels of rugged stone, over which a foaming torrent forces its way in the rainy season, and is succeeded in dry weather by a sparkling rivulet, which trickles down to swell a little brooklet at the foot of the hill, as it winds its way Page 211 to the neighbouring lake. These may be seen, and the patches of heather, and the patient colley watching for a signal to collect the scattered flock, dotted, as it appears to be, over the almost inaccessible heights. At some distance it is difficult to see the sheep, at least by a stranger, partly on account of the dark colour of their fleeces (for they have not the whiteness of our flocks in the midland downs), and partly from the shadow on the hills. Separated as they are from each other, as the evening closes in the sagacious dog receives a hint from his master, and the sheep are quickly collected from places to which the shepherd could with difficulty make his way. Snow and frost are no check to the labours of the colley dog. His exertions are indefatigable, and the only reward he appears to expect is the approbation of his master.

The following amusing anecdote of a sort of sheep-dog was communicated to me by its owner. The dog's name was Hero. His habits were odd enough, and he gave many instances of his sagacity. The following was one of them: —

Hero was in the constant habit of accompanying the farm-horses in their daily labour, pacing the ploughed field regularly aside the team, and returning with them to and from his meals, always taking care to scamper home at a certain hour for a more dainty portion when his mistress dined.

During one of these hasty visits he met a young woman, whom he had never seen before, wearing his Page 212 mistress's cloak. After looking at her with a scrutinising eye, he turned round, and followed her closely, to her great dismay, to a neighbouring village four miles off, where the brother of his mistress lived, and into whose house the woman entered. Probably concluding from this circumstance that she was a privileged person, he returned quietly back again. Had she passed the house, the dog would most probably have seized the cloak, in order to restore it to his mistress.

I trust my readers will begin to feel some interest in this sagacious and useful animal, and I will add one or two more well-authenticated anecdotes of him.

Captain Brown says that his friend, Mr. Peter Macarthur, related to him the following anecdote of a shepherd's dog, which belonged to his grandfather, who at that time resided in the Island of Mull:—Upon one occasion a cow had been missed for some days, and no trace of it could be found; and a shepherd's dog, called Drummer, was also absent. On the second or third day the dog returned, and taking Mr. Macarthur's father by the coat, pulled him towards the door, but he did not follow it; he then went to his grandfather, and pulled him in the same way by the coat, but without being attended to; he next went to one of the men-servants, and tugged him also by the coat. Conceiving at last there was something particular which the dog wanted, they agreed to follow him: this Page 213 seemed to give him great pleasure, and he ran barking and frisking before them, till he led them to a cow-shed, in the middle of a field. There they found the cow fixed by the horns to a beam, from which they immediately extricated her and conducted her home, much exhausted for want of food. It is obvious, that but for the sagacity of this faithful animal she certainly would have died.

Mr. John Cobb, farmer at Tillybirnie, parish of Lethnot, near Brechin, during a severe snow-storm in the year 1798, had gone with his dog, called Cæsar, to a spot on the small stream of Paphry (a tributary of the North Esk), where his sheep on such occasions used to take shelter beneath some lofty and precipitous rocks called Ugly Face, which overhung the stream. While employed in driving them out, an immense avalanche fell from these rocks, and completely buried him and his dog. He found all his endeavours to extricate himself from this fearful situation in vain; and at last, worn out, fell asleep. However, his dog had contrived to work his way out, and returned home next day about noon. The dog, by whining and looking in the faces of the family, and afterwards running to the door, showed that he wished them to follow him; they accordingly did so, accompanied by a number of men provided with spades. He led them to the spot where his master was, and, after scraping away the snow which had fallen from the time he had quitted the spot, he quickly disappeared in the hole by which he had effected Page 214 his escape. They began to dig, and by nightfall they found Mr. Cobb quite benumbed, standing in an upright posture; but as life was not quite extinguished he was rolled in warm blankets, and soon recovered. As may well be conceived, he felt the greatest regard for his preserver, and treated him ever afterwards with much tenderness. The colley lived to a great age, and when he died, his master said it gave him as much pain as the death of a child; and he would have buried him in a coffin, had he not thought that his neighbours would turn it into ridicule.

A gentleman of my acquaintance had a sheep-dog, which was generally kept in a yard by the side of his house in the country. One day a beggar made his way into the yard armed with a stout stick, with which he defended himself from the attacks of the dog, who barked at and attempted to bite him. On the appearance of a servant the dog ceased barking, and watching his opportunity, he got behind the beggar, snatched the stick from his hand, and carried it into the road, where he left it.

A shepherd named Clark, travelling home to Hunt-Law, parish of Minto, near Jedburgh, with some sheep, had occasion to pass through a small village, where he went into a public-house to take a dram with some cronies whom he had met on the road, leaving the sheep in charge of the dog. His friends and he had indulged in a crack for several hours, till he entirely forgot his drove. In the meantime the dog had Page 215 wearied, and determined to take the sheep home himself, a distance of about ten miles. The shepherd, on coming to the spot where he had left the animals, found they were gone, but knowing well that he might depend on the fidelity of his dog, he followed the straight way to Hunt-Law. On coming to a gateway which had interrupted their progress, he perceived the dog and sheep quietly reposing; and had it not been for that bar to their course he would have taken them home. Two miles of their way was by a made road, and the rest through an open moor.

"One of the most interesting anecdotes I have known," says Sir Patrick Walker, who related this anecdote to Captain Brown, and the one which follows, "relates to a sheep-dog. The names of the parties have escaped me just now, but I recollect perfectly that it came from an authentic source. The circumstances were these:—A gentleman sold a considerable flock of sheep to a dealer, which the latter had not hands to drive. The seller, however, told him he had a very intelligent dog, which he would send to assist him to a place about thirty miles off; and that when he reached the end of his journey, he had only to feed the dog, and desire him to go home. The dog accordingly received his orders, and set off with the flock and the drover; but he was absent for so many days that his master began to have serious alarms about him, when one morning, to his great surprise, he found the dog returned with a very large flock of sheep, including Page 216 the whole that he had lately sold. The fact turned out to be, that the drover was so pleased with the colley that he resolved to steal him, and locked him up until the time when he was to leave the country. The dog grew sulky, and made various attempts to escape, and one evening he fortunately succeeded. Whether the brute had discovered the drover's intention, and supposed the sheep were also stolen, it is difficult to say; but by his conduct it looked so, for he immediately went to the field, collected the sheep, and drove them all back to his master."

"A few years ago, when upon a shooting party in the Braes of Ranoch, the dogs were so worn out as to be unfit for travel. Our guide said he knew the shepherd, who had a dog that perhaps might help us. He called, and the young man came with his little black colley, to which, as soon as he had conversed with the guide, he said something in Erse. The dog set off in a sneaking sort of manner up the hill, and, when he showed any degree of keenness, we hastened to follow, lest he should set up the birds; but the lad advised us 'to be canny, as it was time eneuch when Lud came back to tell.' In a short space Lud made his appearance on a knoll, and sat down, and the shepherd said we might go up now, for Lud had found the birds. The dog waited till we were ready, and trotted on at his master's command, who soon cautioned us to be on the alert, for Lud signified we were in the midst of the covey. We immediately found this to be the case, Page 217 and in the course of the day the same thing occurred frequently."

The following anecdote will serve to show the strong affection of the sheep-dog; I will give it in the words of a gentleman who witnessed the fact in the north of England.

"The following instance of canine affection came under my observation at a farm-steading, where I happened to be. A colley belonging to the shepherd on the farm appeared very restless and agitated: she frequently sent forth short howls, and moaned as if in great agony. 'What on earth is the matter with the dog?' I asked. 'Ye see, sur,' said the shepherd, 'au drownt a' her whelps i' the pond the day, and she's busy greeting for them.' Of course, I had no objection to offer to this explanation, but resolved to watch her future operations. She was not long in setting off to the pond and fishing out her offspring. One strong brindled pup she seemed to lament over the most. After looking at it for some time, she again set off at a quick rate to a new house then in the course of erection, and scooped out a deep hole among the rubbish. She then, one by one, deposited the remains of her young in it, and covered them up most carefully. After she had fulfilled this task, she resumed her labours among her woolly charge as usual."

In the winter of the year 1795, as Mr. Boulstead's son, of Great Salkeld, in Cumberland, was attending the sheep of his father upon Great Salkeld Common, Page 218 he had the misfortune to fall and break his leg. He was then at the distance of three miles from home—there was no chance of any person's coming in so unfrequented a place within call, and evening was fast approaching. In this dreadful dilemma, suffering extreme pain from the fracture, and laying upon the damp ground at so dreary a season of the year, his fearful situation suggested to him the following expedient. Folding one of his gloves in his pocket-handkerchief, he fastened it round the neck of the dog, and rather emphatically ordered him 'home.' These dogs, trained so admirably to orders and signals during their attendance upon the flock, are well known to be under the most minute subjection, and to execute the commands of their masters with an alacrity scarcely to be conceived.

Perfectly convinced of some inexplicable disquietude from the situation in which his master lay, he set off at a pace which soon brought him to the house, where he scratched with great violence at the door for immediate admittance. This obtained, the parents were in the utmost alarm and consternation at his appearance, especially when they had examined the handkerchief and its contents. Instantly concluding that some accident had befallen their son, they did not delay a moment to go in search of him. The dog, apparently conscious that the principal part of his duty was yet to be performed, anxiously led the way, and conducted the agitated parents to the spot where their son lay Page 219 overwhelmed with pain, increased by the awful uncertainty of his situation. Happily he was removed just at the close of day; and the necessary assistance being procured, he soon recovered. He was never more pleasingly engaged than when reciting the sagacity and affection of his faithful follower, who then became his constant companion.

Mr. Hawkes, farmer of Halling, returning much intoxicated from Maidstone market, with his dog, when the whole face of the country was covered with snow, mistook his path, and passed over a ditch on his right-hand towards the river; fortunately he was unable to get up the bank, or he must have fallen into the Medway, at nearly high water. Overcome with the liquor, Hawkes fell amongst the snow, in one of the coldest nights ever remembered: turning on his back, he was soon asleep; his dog scratched the snow about him, and then mounted upon the body, rolled himself round, and laid him on his master's bosom, for which his shaggy hide proved a seasonable covering. In this state, with snow falling all the time, the farmer and his dog lay the whole of the night; in the morning, a Mr. Finch, who was out with his gun, perceiving an uncommon appearance, proceeded towards it; at his approach, the dog got off the body, shook the snow from him, and by significant actions encouraged Mr. Finch to advance. Upon wiping the snow from the face, the person was immediately recognised, and was conveyed to the first house, when a pulsation in the heart being evident, the Page 220 necessary means to recover him were employed, and in a short time Hawkes was able to relate his own story. In gratitude for his faithful friend, a silver collar was made for his wearing, and thus inscribed: —

> "In man, true friendship I long strove to find, but missed my aim;
> At length I found it in my dog most kind; man! blush for shame."

The following tale is copied from the "Glasgow Post:" —

"A few days since, while Hector Macalister was on the Aran Hills looking after his sheep, six miles from home or other habitation, his two colley dogs started a rabbit, which ran under a large block of granite. He thrust his arm under the stone, expecting to catch it; but instead of doing so, he removed the supports of the block, which instantly came down on his arm, holding him as fast as a vice. His pain was great; but the pangs he felt were greater when he thought of home, and the death he seemed doomed to die. In this position he lay from ten in the morning till four in the afternoon; when, finding that all his efforts to extricate himself were unavailing, he tried several times, without effect, to get his knife out of his pocket to cut his arm off.

"His only chance now was to send home his dogs, with the view of alarming his friends. After much difficulty, as the faithful creatures were most unwilling to leave him, he succeeded; and Mrs. Macalister, seeing them return alone, took the alarm, and col Page 221 lecting the neighbours, went in search of her husband, led on by the faithful colleys. When they came to the spot, poor Macalister was speechless with crying for assistance. It required five strong men to remove the block from his arm.

"A further instance of reason and self-judgment was shown in the colley, which, having to collect some sheep from the sides of a gorge, through which ran a morass, saw one of the animals precipitate itself into the shifting mass, where it sank immediately up to the neck, leaving nothing but its small black head visible. The dog looked at the sheep and then at its master with an embarrassed, what-shall-I-do kind of expression; but the latter, being too far off to notice the difficulty or to assist, the dog, with infinite address, seized the struggling animal by the neck, and dragged it by main force to the dry land, and then compelled it to join the flock he was collecting."

The care a sheep-dog will take of the sheep committed to his charge is extraordinary, and he will readily chastise any other dog which happens to molest them. Col. Hamilton Smith relates that a strange cur one day bit a sheep in rear of the flock, unseen by the shepherd. The assault was committed by a tailor's dog, but not unnoticed by the other, which immediately seized the delinquent by the ear and dragged him into a puddle, where he kept dabbling him in the mud with the utmost gravity. The cur yelled. The tailor came slipshod with his goose to the rescue, and flung it at Page 222 the sheep-dog, but missed him, and did not venture to pick it up till the castigation was over.

And here I cannot do better than introduce Dr. Walcot's (Peter Pindar) charming lines on "The Old Shepherd's Dog:"—

"The old shepherd's dog, like his master, was grey,
His teeth all departed, and feeble his tongue;
Yet where'er Corin went he was follow'd by Tray:
Thus happy through life did they hobble along.

When fatigued on the grass the shepherd would lie
For a nap in the sun, 'midst his slumbers so sweet
His faithful companion crawl'd constantly nigh,
Placed his head on his lap, or laid down at his feet.

When winter was heard on the hill and the plain,
When torrents descended, and cold was the wind;
If Corin went forth 'mid the tempest and rain,
Tray scorn'd to be left in the chimney behind.

At length, in the straw, Tray made his last bed—
For vain against death is the stoutest endeavour—
To lick Corin's hand he rear'd up his weak head,
Then fell back, closed his eyes, and ah! closed them for ever.

Not long after Tray did the shepherd remain,
Who oft o'er his grave with true sorrow would bend;
And when dying, thus feebly was heard the poor swain,
'O bury me, neighbours, beside my old friend!'"

There can be little doubt but that the dog I have been describing is possessed of almost human sagacity. The following is an extraordinary instance of it. It is related by Dr. Anderson:—

A young farmer in the neighbourhood of Inner Page 223 leithen, whose circumstances were supposed to be good, and who was connected with many of the best store-farming families in the county, had been tempted to commit some extensive depredations upon the flocks of his neighbours, in which he was assisted by his shepherd. The pastoral farms of Tweeddale, which generally consist each of a certain range of hilly ground, had in those days no enclosures: their boundaries were indicated only by the natural features of the country. The sheep were, accordingly, liable to wander, and to become intermixed with each other; and at every reckoning of a flock a certain allowance had to be made for this, as for other contingencies. For some time Mr. William Gibson, tenant in Newby, an extensive farm stretching from the neighbourhood of Peebles to the borders of Selkirkshire, had remarked a surprising increase in the amount of his annual losses. He questioned his shepherds severely, taxed them with carelessness in picking up and bringing home the dead, and plainly intimated that he conceived some unfair dealing to be in progress. The men, finding themselves thus exposed to suspicions of a very painful kind, were as much chagrined as the worthy farmer himself, and kept their minds alive to every circumstance which might tend to afford any elucidation of the mystery. One day, while they were summering their lambs, the eye of a very acute old shepherd, named Hyslop, was caught by a black-faced ewe which they had formerly missed (for the shepherds generally know every parti Page 224 cular member of their flocks), and which was now suckling its own lamb as if it had never been absent. On inspecting it carefully, it was found to bear an additional birn upon its face. Every farmer, it must be mentioned, impresses with a hot iron a particular letter upon the faces of his sheep, as a means of distinguishing his own from those of his neighbours. Mr. Gibson's birn was the letter T, and this was found distinctly enough impressed on the face of the ewe. But above this mark there was an O, which was known to be the mark of the tenant of Wormiston, the individual already mentioned. It was immediately suspected that this and the other missing sheep had been abstracted by that person; a suspicion which derived strength from the reports of the neighbouring shepherds, by whom, it appeared, the black-faced ewe had been tracked for a considerable way in a direction leading from Wormiston to Newby. It was indeed ascertained that instinctive affection for her lamb had led this animal across the Tweed, and over the lofty heights between Cailzie and Newby; a route of very considerable difficulty, and probably quite different from that by which she had been led away, but the most direct that could have been taken. Mr. Gibson only stopped to obtain the concurrence of a neighbouring farmer, whose losses had been equally great, before proceeding with some of the legal authorities to Wormiston, where Millar the shepherd, and his master, were taken into custody, and conducted to the prison Page 225 of Peebles. On a search of the farm, no fewer than thirty-three score of sheep belonging to various individuals were found, all bearing the condemnatory O above the original birns; and it was remarked that there was not a single ewe returned to Grieston, the farm on the opposite bank of the Tweed, which did not minny her lambs—that is, assume the character of mother towards the offspring from which she had been separated.

The magnitude of this crime, the rareness of such offences in the district, and the station in life of at least one of the offenders, produced a great sensation in Tweeddale, and caused the elicitation of every minute circumstance that could possibly be discovered respecting the means which had been employed for carrying on such an extensive system of depredation. The most surprising part of the tale is the extent to which it appears that the instinct of dumb animals had been instrumental, both in the crime and in its detection. While the farmer seemed to have deputed the business chiefly to his shepherd, the shepherd seemed to have deputed it again, in many instances, to a dog of extraordinary sagacity, which served him in his customary and lawful business. This animal, which bore the name of "Yarrow," would not only act under his immediate direction in cutting off a portion of a flock, and bringing it home to Wormiston, but is said to have been able to proceed solitarily, and by night, to a sheepwalk, and there detach certain individuals previously Page 226 pointed out by its master, which it would drive home by secret ways, without allowing one to straggle. It is mentioned that, while returning home with their stolen droves, they avoided, even in the night, the roads along the banks of the river, or those that descend to the valley through the adjoining glens. They chose rather to come along the ridge of mountains that separate the small river Leithen from the Tweed. But even here there was sometimes danger, for the shepherds occasionally visit their flocks even before day; and often when Millar had driven his prey from a distance, and while he was yet miles from home, and the weather-gleam of the eastern hills began to be tinged with the brightening dawn, he has left them to the charge of his dog, and descended himself to the banks of the Leithen, off his way, that he might not be seen connected with their company. Yarrow, although between three and four miles from his master, would continue, with care and silence, to bring the sheep onward to Wormiston, where his master's appearance could be neither a matter of question nor surprise.

Near to the thatched farmhouse was one of those old square towers, or peel-houses, whose picturesque ruins were then seen ornamenting the course of the Tweed, as they had been placed alternately along the north and south bank, generally from three to six hundred yards from it—sometimes on the shin, and sometimes in the hollow of a hill. In the vault of this tower it was the practice of these men to conceal the Page 227 sheep they had recently stolen; and while the rest of their people were absent on Sunday at the church, they used to employ themselves in cancelling with their knives the ear-marks, and impressing with a hot iron a large O upon the face, that covered both sides of the animal's nose, for the purpose of obliterating the brand of the true owner. While his accomplices were so busied, Yarrow kept watch in the open air, and gave notice, without fail, by his barking, of the approach of strangers.

The farmer and his servant were tried at Edinburgh in January 1773, and the proceedings excited an extraordinary interest, not only in the audience, but amongst the legal officials. Hyslop, the principal witness, gave so many curious particulars respecting the instincts of sheep, and the modes of distinguishing them both by natural and artificial marks, that he was highly complimented by the bench. The evidence was so complete, that both culprits were found guilty and expiated their crime on the scaffold.

The general tradition is, that Yarrow was also put to death, though in a less ceremonious manner; but this has probably no other foundation than a *jeu d'esprit*, which was cried through the streets of Edinburgh as his dying speech. We have been informed that the dog was in reality purchased, after the execution of Millar, by a sheep-farmer in the neighbourhood, but did not take kindly to honest courses, and his new master having no work of a different kind in which to Page 228 engage him, he was remarked to show rather less sagacity than the ordinary shepherd's dog.

An instance of shrewd discrimination in the shepherd's dog, almost as remarkable as that of poor Yarrow, was mentioned a few years ago in a Greenock newspaper. In the course of last summer, says the narrator, it chanced that the sheep on the farm of a friend of ours, on the water of Stinchar, were, like those of his neighbours, partially affected with that common disease, maggots in the skin, to cure which distemper it is necessary to cut off the wool over the part affected, and apply a small quantity of tobacco juice, or some other liquid. For this purpose the shepherd set off to the hill one morning, accompanied by his faithful canine assistant, Ladie. Arrived among the flock, the shepherd pointed out a diseased animal; and making the accustomed signal for the dog to capture it, "poor Mailie" was speedily sprawling on her back, and gently held down by the dog till the arrival of her keeper, who proceeded to clip off a portion of her wool, and apply the healing balsam. During the operation, Ladie continued to gaze on the operator with close attention; and the sheep having been released, he was directed to capture in succession two or three more of the flock, which underwent similar treatment. The sagacious animal had now become initiated into the mysteries of his master's vocation, for off he set unbidden through the flock, and picked out with unerring precision those sheep which were affected with mag Page 229 gots in their skin, and held them down until the arrival of his master; who was thus, by the extraordinary instinct of Ladie, saved a world of trouble, while the operation of clipping and smearing was also greatly facilitated.

Often as I have attempted to make acquaintance with a colley-dog, I have never been able to succeed in producing any degree of familiarity. On the contrary, he has always regarded me with looks of shyness and suspicion. His master appears to be the only being to whom he is capable of showing any degree of attachment; and coiled up on his great-coat, or reposing at his feet, he eyes a stranger with distrust, if not with anger. At the same time there is a look of extraordinary intelligence, which perhaps is possessed by no other animal in a greater degree. It has been said of him, that although he has not the noble port of the Newfoundland dog, the affectionate fondling of the spaniel, nor the fierce attachment which renders the mastiff so efficient a guard, yet he exceeds them all in readiness and extent of intelligence, combined with a degree of docility unequalled, perhaps, by any other animal in existence. There is, if the expression may be used, a philosophic look about him, which shows thought, patience, energy, and vigilance. During a recent visit in Cumberland, I took some pains to make myself acquainted with the character of this dog, and I am now convinced that too much cannot be said of his wonderful properties. He protects with indefatigable Page 230 exertions the flock committed to his charge. When we consider the dreary wilds, the almost inaccessible heights, the rugged hills and lofty mountains to which sheep have access, and to which man could scarcely penetrate—that some sheep will stray and intermix with other flocks—that the dog knows the extent of his walk as well as every individual of his flock, and that he will select his own as well as drive away intruders, we must admit his utility and admire his sagacity.

Let me give another instance of this in the words of the Ettrick Shepherd. It was related to me by himself, and has since been published in the "Percy Anecdotes."

"I once witnessed a very singular feat performed by a dog belonging to John Graham, late tenant in Ashiesteel. A neighbour came to his house after it was dark, and told him that he had lost a sheep on his farm, and that if he (Graham) did not secure her in the morning early, she would be lost, as he had brought her far. John said he could not possibly get to the hill next morning, but if he would take him to the very spot where he lost the sheep, perhaps his dog Chieftain would find her that night. On that they went away with all expedition, lest the traces of the feet should cool; and I, then a boy, being in the house, went with them. The night was pitch dark, which had been the cause of the man losing his ewe, and at length he pointed out a place to John by the Page 231 side of the water where he had lost her. 'Chieftain, fetch that!' said John. 'Bring her back, sir!' The dog jumped around and around, and reared himself up on end; but not being able to see anything, evidently misapprehended his master, on which John fell to scolding his dog, calling it a great many hard names. He at last told the man that he must point out the very track that the sheep went, otherwise he had no chance of recovering it. The man led him to a grey stone, and said he was sure she took the brae (hill side) within a yard of that. 'Chieftain, come hither to my foot, you great numb'd whelp!' said John. Chieftain came—John pointed with his finger to the ground, 'Fetch that, I say, sir—bring that back—away!' The dog scented slowly about on the ground for some seconds, but soon began to mend his pace, and vanished in the darkness. 'Bring her back!—away, you great calf!' vociferated John, with a voice of exultation, as the dog broke to the hill; and as all these good dogs perform their work in perfect silence, we neither saw nor heard any more of him for a long time. I think, if I remember right, we waited there about half an hour, during which time all the conversation was about the small chance which the dog had to find the ewe, for it was agreed on all hands that she must long ago have mixed with the rest of the sheep on the farm. How that was, no man will ever be able to decide. John, however, still persisted in waiting until his dog came back, either with the ewe or Page 232 without her. At last the trusty animal brought the individual lost sheep to our very feet, which the man took on his back, and went on his way rejoicing."

The care the shepherds of the north of England take in preserving a pure breed of these dogs is very great, and the value set upon them is proportionably high. Nor must the shepherds themselves be passed over without notice. They are a shrewd, sagacious set of men, many of them by no means uneducated, as is the case generally with the peasantry in the north of England. Indeed, it is from this class that many scholars and mathematicians have done so much credit, and I may add honour, to the counties of Cumberland and Westmoreland. An anecdote is related of a shepherd, who was found by a gentleman attending his flock, and reading a volume of Milton. "What are you reading?" asked the gentleman. "Why," replied the shepherd, "I am reading an odd sort of a poet; he would fain rhyme, but does not quite know how to set about it."

The valleys, or glens, which intersect the Grampian mountains, are chiefly inhabited by shepherds. The pastures over which each flock is permitted to range extend many miles in every direction. The shepherd never has a view of his whole flock at once, except when they are collected for sale or shearing. His occupation is to make daily excursions to the different extremities of his pastures in succession, and to turn back, by means of his dog, any stragglers that may be Page 233 approaching the boundaries of his neighbours. In one of these excursions, a shepherd happened to carry along with him one of his children, about three years old. This is a usual practice among the Highlanders, who accustom their children from their earliest infancy to endure the rigours of the climate. After traversing his pasture for some time, attended by his dog, the shepherd found himself under the necessity of ascending a summit at some distance, in order to have a more extensive view of his range. As the ascent was too fatiguing for the child, he left him on a small plain at the bottom, with strict injunctions not to stir from it till his return. Scarcely, however, had he gained the summit, when the horizon was suddenly darkened by one of those impenetrable mists which frequently descend so rapidly amidst these mountains, as almost to turn day into night, and that in the course of a few minutes. The anxious father instantly hastened back to find his child, but, owing to the unusual darkness, he missed his way in the descent. After a search of many hours amongst the dangerous morasses and cataracts with which these mountains abound, he was at length overtaken by night. Still wandering on without knowing whither, he at length came to the verge of the mist, and, by the light of the moon, discovered that he had reached the bottom of his valley, and was within a short distance of his cottage. To renew the search that night was equally fruitless and dangerous. He was, therefore, obliged to return to his Page 234 cottage, having lost both his child and his dog, who had attended him faithfully for years.

Next morning by daybreak, the shepherd, accompanied by a band of his neighbours, set out in search of the child, but, after a day spent in fruitless fatigue, he was at last compelled, by the approach of night, to descend from the mountain. On returning to his cottage he found that the dog, which he had lost the day before, had been home, and on receiving a piece of cake, had instantly gone off again. For several successive days the shepherd renewed the search for his child, but still, on returning at evening disappointed to his cottage, he found that the dog had been home, and, on receiving his usual allowance of cake, had instantly disappeared. Struck with this circumstance, he remained at home one day, and when the dog, as usual, departed with his piece of cake, he resolved to follow him, and find out the cause of his strange procedure. The dog led the way to a cataract, at some distance from the spot where the shepherd had left his child. The banks of the cataract almost joined at the top, yet separated by an abyss of immense depth, presenting that appearance which so often astonishes and appals travellers who frequent the Grampian Mountains, and indicates that these stupendous chasms were not the silent work of time, but the sudden effect of some violent convulsion of the earth. Down one of these rugged and almost perpendicular descents, the dog began, without hesitation, to make his way, and at last Page 235 disappeared into a cave, the mouth of which was almost on a level with the torrent. The shepherd with some difficulty followed, but upon entering the cave, what were his emotions when he beheld his lost child eating with much satisfaction the cake which the dog had just brought to him, while the faithful animal stood by, eyeing his y-oung charge with the utmost complacence.

From the situation in which the child was found, it appears that he had wandered to the brink of the precipice, and then either fallen or scrambled down till he reached the cave, which the dread of the torrent had probably prevented him from quitting. The dog had traced him to the spot, and afterwards prevented him from starving by giving up to him the whole, or the greater part of his own daily allowance. He appears never to have quitted the child by night or day, except when it was necessary to go for food, and then he was always seen running at full speed to and from the cottage.

This extraordinary and interesting anecdote is taken from the "Monthly Magazine" of April, 1802, and bears every appearance of authenticity. It affords an instance of the sense, affection, and self-denial of a faithful animal, and is recorded to his honour, and as an example to the whole race of human beings.

Mr. Daniel, in the Supplement to his "Rural Sports," gives the following account of the shepherds' dogs in Page 236 North Wales. He says, "The sheep in this country are the ancient Alpine sort, (how excellent the mutton is!) and that from their varying mode of life they assume very different habits to the sheep of an inland country, while those of the shepherds' dogs are no less conspicuous. The excellency of these animals renders sheep-pens in a great degree unnecessary. If a shepherd wishes to inspect his flock in a cursory way, he places himself in the middle of the field, or the piece of ground they are depasturing, and giving a whistle or a shout, the dogs and the sheep are equally obedient to the sound, and draw towards the shepherd, and are kept within reach by one or more dogs, until the business which required them to be assembled is finished. In such estimation was this breed of dogs, when cattle constituted one of the grand sources of wealth to the country, that in the laws of Hywell Dda, the legal price of one perfectly broken in for conducting the flocks or herds to or from their pasturage, was equal to that of an ox, viz. sixty denarii, while the price of the house-dog was estimated at only four, which was the value of a sheep. If any doubt arose as to the genuineness of the breed, or his having been *pastorally* trained, then the owner and a neighbour were to make oath that he went with the flocks or herds in the morning, and drove them, with the stragglers, home in the evening."

I delight in seeing a shepherd's dog in full activity, Page 237 anxious to obey the directions of his master. He runs with his utmost speed, encompassing a large space of open country in a short time, and brings those sheep that are wanted to the feet of his master. Indeed the natural talents and sagacity of this dog are so great, partly by being the constant companion of his master, and partly by education, that he may almost be considered a rational being. Mr.

Smellie says, "that he reigns at the head of his flock, and that his *language*, whether expressive of blandishment or of command, is better heard and better understood than the voice of his master. Safety, order, and discipline are the effects of his vigilance and activity. Sheep and cattle are his subjects. These he conducts and protects with prudence and bravery, and never employs force against them, except for the preservation of peace and good order. He not only understands the language of his master, but, when too distant to be heard, he knows how to act by signals made with the hand." How well Delille describes this faithful animal! –

> "Aimable autant qu'utile,
> Superbe et caressant, courageux et docile,
> Formé pour le conduire et pour le protéger.
> Du troupeau qu'il gouverne il est le vrai berger;
> Le Ciel l'a fait pour nous; et dans leur cours rustique,
> Il fut des rois pasteurs le premier domestique."

Mr. Charles Darwin, in his interesting travels in South America, informs us, that when riding it is a Page 238 common thing to meet a large flock of sheep, guarded by one or two dogs, at the distance of some miles from any house or man. He often wondered how so firm a friendship had been established, till he found that the method of education consisted in separating the puppy, while very young, from the mother, and in accustoming it to its future companions. In order to do this, a ewe is held three or four times a-day for the little thing to suck, and a nest of wool is made for it in the sheep-pen. At no time is it allowed to associate with other dogs, or with the children of the family. From this education, it has no wish to leave the flock, and just as another dog will defend his master, so will these the sheep. It is amusing to observe, when approaching a flock, how the dog immediately advances barking, and the sheep all close in his rear, as if round the oldest ram. These dogs are also easily taught to bring home the flock at a certain hour in the evening. Their most troublesome fault, when young, is their desire of playing with the sheep; for, in their sport, they sometimes gallop their poor subjects most unmercifully. The shepherd dog comes to the house every day for some meat, and immediately it is given him he skulks away as if ashamed of himself. On these occasions the house-dogs are very tyrannical, and the least of them will attack and pursue the stranger. The minute, however, the latter has reached the flock, he turns round and begins to bark, and then all the house-dogs take very Page 239 quietly to their heels. In a similar manner, a whole pack of hungry wild dogs will scarcely ever venture to attack a flock when under the protection of even one of these faithful shepherds.

Page 240

THE ST. BERNARD DOG.

"Thrill sounds are breaking o'er the startled ear,
The shriek of agony, the cry of fear;—
And the sad tones of childhood in distress,
Are echoing through the snow-clad wilderness!

And who the first to waken to the sound,
And quickly down the icy path to bound;
To dare the storm with anxious step and grave,
The first to answer and the first to save?—
'T is he—the brave old dog, who many a day
Hath saved lost wand'rers in that dreary way;
And now, with head close crouched along the ground,
Is watching eagerly each coming sound.
Sudden he starts—the cry is near—
On, gallant Bruno!—know no fear!
On!—for that cry may be the last,
And human life is ebbing fast! Page 241
And now he hurries on with heaving side,
Dashing the snow from off its shaggy hide;—
He nears the child!—he hears his gasping sighs,
And, with a tender care, he bears away the prize."—Mrs. Houstoun.

Sir Walter Scott said that he would believe anything of a St. Bernard dog. Their natural sagacity is, indeed, so sharpened by long practice and careful training, that a sort of language is established between them and the good monks of St. Bernard, by which mutual communications are made, such as few persons living in situations of less constant and severe trials can have any just conceptions of. When we look at the extraordinary sagacity of the animal, his great strength, and his instinctive faculties, we shall feel convinced how admirably he is adapted to fulfil the purpose for which he is chiefly employed,—that of saving lives in snow-storms.

The peculiar faculty of the St. Bernard dogs is shown by the curious fact, that if a whelp of this breed is placed upon snow for the first time, it will begin to scratch it, and sniff about as if in search of something. When they have been regularly trained, they are generally sent out in pairs during heavy snow-storms in search of travellers, who may have been overwhelmed by the snow. In this way they pass over a great extent of country, and by the acuteness of their scent discover if any one is buried in the snowdrift. When it is considered that Mount St. Bernard is situated about 8000 feet above the level of the sea, and that it is the highest Page 242 habitable spot in Europe, and that the road which passes across it is constantly traversed, the great utility of the dogs is sufficiently manifest. Neither is the kindness, charity, and hospitality of the good monks less to be admired than the noble qualities of these dogs.

"Under every circumstance," says Mr. Brockedon, "in which it is possible to render assistance, the worthy religieuses of St. Bernard set out upon their fearful duty unawed by the storm, and obeying a higher Power; they seek the exhausted or overwhelmed traveller, accompanied by their dogs, whose sagacity will generally detect the victim though buried in the snow. The dogs, also, as if conscious of a high duty, will roam alone through the day and night in these desolate regions, and if they discover an exhausted traveller will lie on him to impart warmth, and bark and howl for assistance." [P]

Mr. Mathews, in his "Diary of an Invalid," gives this testimony in praise of the inmates of St. Bernard. "The approach," he says, "to the convent for the last Page 243 hour of the ascent is steep and difficult. The convent is not seen till you arrive within a few hundred yards of it; when it breaks upon the view all at once, at a turn in the rock. Upon a projecting crag near it stood one of the celebrated dogs, baying at our advance, as if to give notice of strangers. These dogs are of a large size, particularly high upon the legs, and generally of a milk white, or of a tabby colour. They are most extraordinary creatures, if all the stories the monks tell of them are true. They are used for the purpose of searching for travellers who may be buried in the snow; and many persons are rescued annually from death by their means. During the last winter, a traveller arrived at the convent in the midst of a snow-storm, having been compelled to leave his wife, who was unable to proceed further, at about a quarter of a mile's distance. A party of the monks immediately set out to her assistance, and found her completely buried under the snow. The sagacity of the dogs alone was the cause of her deliverance, for there was no visible trace, and it is difficult to understand how the scent can be conveyed through a deep covering of snow.

"It is stated that the monks themselves, when out upon search for travellers, have frequently owed their preservation to their dogs, in a manner which would seem to show that the dogs are endued with a presentiment of danger.

"Many stories of this kind have been told, and I was anxious to ascertain their truth. The monks Page 244 stated two or three cases where the dogs had actually prevented them from returning to the convent by their accustomed route, when it afterwards turned out, that if they had not followed the guidance of their dog in his deviation, they would have been overwhelmed by an avalanche. Whether the dog may be endued with an intuitive foreboding of danger, or whether he may have the faculty of detecting symptoms not perceptible to our duller senses, must be determined by philosophers."

That dogs and other animals, especially elephants, have this faculty, cannot be doubted. There is an instance on record of a dog having, by his importunity and peculiar gestures, induced his mistress to quit a washhouse in which she was at work, the roof of which fell in almost immediately afterwards. Dogs have been known to give the alarm of fire, by howling and other signs, before it was perceived by any of the inmates of the house. Their apprehension of danger is indeed very acute and very extraordinary, and may serve to account for and prove the accuracy of what has been stated respecting the instinct of the St. Bernard dogs.

These dogs, however, do not always escape being overwhelmed by a sudden avalanche, which falls, as is most usual, in the spring of the year. Two of the domestics of the convent, with two or three dogs, were escorting some travellers, and were lost in an avalanche. One of the predecessors of these dogs, an intelligent animal, which had served the hospital for the space of Page 245 twelve years, had, during that time, saved the lives of many individuals. Whenever the mountain was enveloped in fogs and snow, he set out in search of lost travellers. He was accustomed to run barking until he lost his breath, and would frequently venture on the most perilous places. When he found his strength was insufficient to draw from the snow a traveller benumbed with cold, he would run back to the hospital in search of the monks.

One day this interesting animal found a child in a frozen state between the Bridge of Drouaz and the Ice-house of Balsora. He immediately began to lick him, and having succeeded in restoring animation, and the perfect recovery of the boy, by means of his caresses, he induced the child to tie himself round his body. In this way he carried the poor little creature, as if in triumph, to the hospital. When old age deprived him of strength, the prior of the convent pensioned him at Berne by way of reward. He is now dead, and his body stuffed and deposited in the museum of that town. The little phial, in which he carried a reviving liquor for the distressed travellers whom he found among the mountains, is still suspended from his neck.

The story of this dog has been often told, but it cannot be too frequently repeated. Its authenticity is well established, and it affords another proof of the utility and sense of the St. Bernard dogs. Neither can the benevolence of the good monks be too highly praised. To those accustomed to behold the habitations Page 246 of man, surrounded by flowery gardens, green and pleasing meadows, rivulets winding and sparkling over their pebbly bottoms, and groves in which songsters haunt and warble, the sight of a large monastery, situated on a gigantic eminence, with clouds rolling at its foot, and encompassed only by beds of ice and snow, must be awfully impressive. Yet amidst these boundless labyrinths of rugged glens and precipices, in the very rudest seasons, as often as it snows or the weather is foggy, do some of those benevolent persons go forth, with long poles, guided by their sagacious dogs. In this way they seek the high road, which these animals, with their instinctive faculty, never miss, how difficult soever to find. If an unfortunate traveller has sunk beneath the force of the falling snows, or should be immersed among them, the dogs never fail to find the place of his interment, which they point out by scratching and snuffing; when the sufferer is dug out, and carried to the monastery, where means are used for his recovery.

The Count de Monte Veccios had a St. Bernard dog, which, as his master always had reported, could understand whatever he said to him; and the following short account deserves to be recorded, as it at once indicates memory, compassion, love, gratitude, and resentment in the faithful animal, even if we do not allow it to make good his master's opinion. The story is this: —

The Count had served long in the wars, and always Page 247 had this faithful attendant with him. The republic of Venice had been signally indebted to his courage, but had not rewarded him. He had a favour to ask of the then General Morosini; and as that commander was a man of singular pride and arrogance, he was obliged to wait a favourable opportunity of presenting his suit. One day when the General himself had a favour to ask of the Doge (who was a person of high elegance, and celebrated for his love of expensive entertainments), he laid out half his fortune on a cold collation, to which he had invited the Doge, to put him in humour for his suit. Thinking this the most suitable time for his purpose, as he who was about to ask a favour for himself would hardly at that instant deny one to another, the Count went to him some hours before the Doge was expected, and was graciously received in the room where the table was prepared. Here he began to make his court to the General, by praising the elegance and pomp of the preparation, which consisted of many thousands of finely-cut vessels of Venetian glass, filled with the richest sweetmeats and cold provisions, and disposed on fine tables, all covered with one vast cloth, with a deep gold fringe, which swept the ground. The Count said a thousand fine things about the elegance and richness of the dessert, and particularly admired the profusion of expense in the workmanship of the crystal and the weight of the gold fringe. Thus far he was very courteously treated; and the lord of the feast pompously Page 248 told him that all the workmen in Venice had been half a year employed about them. From this he proceeded to the business of his suit; but this met with a very different reception, and was not only refused, but the denial attended with very harsh language. The Count was shocked at the ill-nature of the General, and went away in a very melancholy mood. As he went out, he patted his dog upon the head, and, out of the fulness of his heart, said to him with an afflicted air, "*Tu vois, mon ami, comme l'on nous traite*,—You see, my friend, how I am used." The dog looked up wistfully in his face, and returned him an answer with his tears. He accompanied him till he was at some distance from the General's, when, finding him engaged in company, he took that opportunity of leaving him with people who might justify him if accused. Upon which the dog, returning back to the house of the haughty officer, entered the great room, and taking hold of the gold tassel at one of the corners of the cloth, ran forcibly back, and drew after him the whole preparation, which in a moment lay strewed on the ground in a vast heap of broken glasses; thus revenging his master's quarrel, and ensuring as unexpected a reception to the General's requests as the latter had given to those of the Count.

One of the St. Bernard dogs, named Barry, had a medal tied round his neck as a badge of honourable distinction, for he had saved the lives of forty persons. He at length died nobly in his vocation. In the winter Page 249 of 1816, a Piedmontese courier arrived at St. Bernard on a very stormy day, labouring to make his way to the little village of St. Pierre, in the valley beneath the mountain, where his wife and children lived. It was in vain that the monks attempted to check his resolution to reach his family. They at last gave him two guides, each of whom was accompanied by a dog, one of which was the remarkable creature whose services had been so valuable. They set forth on their way down the mountain. In the mean time the anxious family of the poor courier, alarmed at his long absence, commenced the ascent of the mountain, in hopes of meeting him, or obtaining some information respecting him. Thus at the moment he and his guides were descending, his family were toiling up the icy steep, crowned with the snows of ages. A sudden crackling noise was heard, and then a thundering roar echoing through the Alpine heights—and all was still. Courier, and guides, and dogs, and the courier's family, were at the same moment overwhelmed by one common destruction—not one escaped. Two avalanches had broken away from the mountain pinnacles, and swept with impetuous force into the valley below.

Page 250

CHASSEUR AND CUBA BLOODHOUNDS.

THE BLOODHOUND.

"His snuffling nose, his active tail,
Attest his joy; then with deep op'ning mouth,
That makes the welkin tremble, he proclaims
Th' audacious felon; foot by foot he marks

His winding way, while all the listening crowd
Applaud his reasonings. O'er the watery ford,
Dry sandy heaths, and stony barren hills,
O'er beaten paths, with men and beasts distain'd,
Unerring he pursues; till at the cot
Arriv'd, and seizing by his guilty throat
The caitiff vile, redeems the captive prey:
So exquisitely delicate his sense!"—Somerville.

Page 251 These noble dogs were also called "Slough dogs," in consequence of their exploring the sloughs, mosses, and bogs, in pursuit of offenders, called Moss-troopers. They were used for this purpose as late as the reign of James the First. In Scotland they are called the Sleuth-hound. It is the largest of any variety of hound, some of them having measured from twenty-six to twenty-eight inches to the top of the shoulder. They are beautifully formed, and have a noble expression of countenance, so finely portrayed in Sir Edwin Landseer's well-known and beautiful picture of "Dignity and Impudence." There is, as Colonel Hamilton Smith has observed, a kind of sagacious, or serious, solemn dignity about him, admirably calculated to impress the marauder with dread and awe. Indeed, so much is this the case, that I knew an instance of a bloodhound having traced a sheep-stealer to his cottage in Bedfordshire; and so great was the dread afterwards of the peculiar instinct of this dog, that sheep-stealing, which had before been very common in the neighbourhood, was put an end to. It has, therefore, often occurred to me, that if bloodhounds were kept for the general good in different districts, sheep-stealing would be less frequent than it is at present. They Page 252 might also be usefully employed in the detection of rick-burners. At all events the suggestion is worth some consideration, especially from insurance offices. In 1803, the Thrapston Association for the Prosecution of Felons in Northamptonshire, procured and trained a bloodhound for the detection of sheep-stealers. In order to prove the utility of the dog, a man was dispatched from a spot where a great concourse of people were assembled, at ten o'clock in the forenoon, and an hour afterwards the hound was laid on the scent. After a chase of an hour and a half, the hound found him secreted in a tree many miles from the place of starting. The very knowledge that farmers could readily have recourse to the assistance of such a dog, would serve to prevent the commission of much crime.

To try whether a young bloodhound was well instructed, a nobleman (says Mr. Boyle) caused one of his servants to walk to a town four miles off, and then to a market-town three miles from thence. The dog, without seeing the man he was to pursue, followed him by the scent to the above-mentioned places, notwithstanding the multitude of people going the same road, and of travellers that had occasion to cross it. When the hound came to the chief market-town, he passed through the streets, without noticing any of the people there, till he got to the house where the man he sought was, and there found him in an upper room.

A sure way of stopping the dog was to spill Page 253 blood upon the track, which destroyed the discriminating fineness of his scent. A captive was sometimes sacrificed on such occasions. Henry the Minstrel tells us a romantic story of Wallace, founded on this circumstance. The hero's little band had been joined by an Irishman named Fawdon, or Fadzean, a dark, savage, and suspicious character. After a sharp skirmish at Black Erneside, Wallace was forced to retreat with only sixteen followers. The English pursued with a border sleuth-bratch, or bloodhound. In the retreat, Fawdon, tired, or affecting to be so, would go no farther. Wallace having in vain argued with him, in hasty anger struck off his head, and continued the retreat. When the English came up, their hound stayed upon the dead body.

To the present group has been referred by some naturalists a dog of Spanish descent, termed the Cuban bloodhound. A hundred of these sagacious but savage dogs were sent, in 1795, from the Havanna to Jamaica, to extinguish the Maroon war, which at that time was fiercely raging. They were accompanied by forty Spanish chasseurs, chiefly people of colour, and their appearance and that of the dogs struck terror into the negroes. The dogs, muzzled and led in leashes, rushed ferociously upon every object, dragging along the chasseurs in spite of all their endeavours. Dallas, in his "History of the Maroons," informs us that General Walpole ordered a review of these dogs and the men, that he might see in what manner they Page 254 would act. He set out for a place called Seven Rivers, accompanied by Colonel Skinner, whom he appointed to conduct the attack. "Notice of his coming having preceded him, a parade of the chasseurs was ordered, and they were taken to a distance from the house, in order to be advanced when the general alighted. On his arrival, the commissioner (who had procured the dogs), having paid his respects, was desired to parade them. The Spaniards soon appeared at the end of a gentle acclivity drawn out in a line, containing upwards of forty men, with their dogs in front unmuzzled, and held by cotton ropes. On receiving the command, 'Fire!' they discharged their fusils, and advanced as upon a real attack. This was intended to ascertain what effect would be produced on the dogs if engaged under a fire of the Maroons. The volley was no sooner discharged than the dogs rushed forward with the greatest fury, amid the shouts of the Spaniards, who were dragged on by them with irresistible force. Some of the dogs, maddened by the shout of attack while held back by the ropes, seized on the stocks of the guns in the hands of their keepers, and tore pieces out of them. Their impetuosity was so great that they were with difficulty stopped before they reached the general, who found it necessary to get expeditiously into the chaise from which he had alighted; and if the most strenuous exertions had not been made, they would have seized upon his horses." This terrible exhibition produced Page 255 the intended effect—the Maroons at once capitulated, and were subsequently sent to Halifax, North America.

Mr. John Lawrence, says that a servant, discharged by a sporting country gentleman, broke into his stables by night, and cut off the ears and tail of a favourite hunter. As soon as it was discovered, a bloodhound was brought into the stable, who at once detected the scent of the miscreant, and traced it more than twenty miles. He then stopped at a door, whence no power could move him. Being at length admitted, he ran to the top of the house, and, bursting open the door of a garret, found the object that he sought in bed, and would have torn him to pieces, had not the huntsman, who had followed him on a fleet horse, rushed up after him.

Colonel Hamilton Smith says, that he was favoured with the following interesting notice of this dog from Sir Walter Scott, and which agrees exactly with some I have seen bred by Lord Bagot at

Blithfield in Staffordshire, and some belonging to Her present Majesty.

"The only sleuth-hound I ever saw was one which was kept at Keeldar Castle. He was like the Spanish pointer, but much stronger, and untameably fierce,—colour, black and tawny, long pendulous ears,—had a deep back, broad nostrils, and was strongly made, Page 256 something like the old English mastiff, now so rare."

Wanley, in his "Wonders of the Little World," relates the following anecdote:—

"Anno Dom. 867.—Lothbroke, of the blood-royal of Denmark, and father to Humbar and Hubba, entered with his hawk into a boat alone, and by tempest was driven upon the coast of Norfolk in England; where being found, he was detained, and presented to Edmund, at that time King of the East Angles. The king entertained him at his court; and perceiving his singular dexterity and activity in hawking and hunting, bore him particular favour. By this means he fell into the envy of Berick, the king's falconer, who one day, as they hunted together, privately murdered and threw him into a bush. It was not long before he was missed at court. When no tidings could be heard of him, his dog, who had continued in the wood with the corpse of his master, till famine forced him thence, at sundry times came to court, and fawned on the king; so that the king, suspecting some ill matter, at length followed the trace of the hound, and was led by him to the place where Lothbroke lay. Inquisition was made; and by circumstance of words, and other suspicions, Berick, the king's falconer, was pronounced to be his murderer. The king commanded him to be set alone in Lothbroke's boat, and committed to the mercy of the sea, by the working of which he was carried to Page 257 the same coast of Denmark from whence Lothbroke came. The boat was well known, and the occupant, Berick, examined by torments. To save himself, he asserted that Lothbroke had been slain by King Edmund. And this was the first occasion of the Danes' arrival in this land."

A planter had fixed his residence at the foot of the Blue Mountains, in the back settlements of America. One day the youngest of his family, a child of about four years old, disappeared. The father, becoming alarmed, explored the woods in every direction, but without success. On the following day the search was renewed, during which a native Indian happened to pass, accompanied by his dog, one of the true bloodhound breed. Being informed of the distress of the planter, he requested that the shoes and stockings last worn by the child might be brought to him. He made the dog smell to them, and patted him. The intelligent animal seemed to comprehend all about it, for he began immediately to sniff around. The Indian and his dog then plunged into the wood. They had not been there long before the dog began to bay; he thought that he had hit upon the scent, and presently afterwards, being assured of it, he uttered a louder and more expressive note, and darted off at full speed into the forest. The Indian followed, and after a considerable time met his dog bounding back, his noble countenance beaming with animation. The hound turned again into the wood, his master not being far behind, Page 258 and they found the child lying at the foot of a tree, fatigued and exhausted, but otherwise unhurt.

Some of these dogs are kept by the keepers in the royal parks and forests, and are used to trace wounded deer. An officer in the 1st Life Guards has two noble dogs of this description, for one of which, I am informed, he gave fifty pounds. In fact, they are by no means uncommon in England. One distinguishing trait of purity in the breed is the colour, which is almost invariably a reddish tan, progressively darkening to the upper part, with a mixture of black upon the back.

"In the Spanish West India Islands," says Bingley, "there are officers called chasseurs, kept in continual employment. The business of these men is to traverse the country with their dogs, for the purpose of pursuing and taking up all persons guilty of murder, or other crimes; and no activity on the part of the offenders will enable them to escape. The following is a very remarkable instance, which happened not many years ago.

"A fleet from Jamaica, under convoy to Great Britain, passing through the Gulf of Mexico, beat upon the north side of Cuba. One of the ships, manned with foreigners (chiefly renegado Spaniards), in standing in with the land at night, was run on shore. The officers, and the few British seamen on board, were murdered, and the vessel was plundered by Page 259 the renegadoes. The part of the coast on which the vessel was stranded being wild and unfrequented, the assassins retired with their booty to the mountains, intending to penetrate through the woods to some remote settlements on the southern side, where they hoped to secure themselves, and elude all pursuit. Early intelligence of the crime had, however, been conveyed to Havanna. The assassins were pursued by a detachment of the Chasseurs del Rey, with their dogs; and in the course of a very few days they were every one apprehended and brought to justice.

"The dogs carried out by the Chasseurs del Rey are all perfectly broken in. On coming up with the fugitive, they bark at him till he stops; they then crouch near him, terrifying him with a ferocious growling if he attempts to stir. In this position they continue barking, to give notice to the chasseurs, who come up and secure their prisoner.

"Each chasseur can only hunt with two dogs. These people live with their dogs, and are inseparable from them. At home the animals are kept chained; and when walking out with their masters, they are never unmuzzled nor let out of ropes, but for attack.

"Bloodhounds were formerly used in certain districts lying between England and Scotland, that were much infested by robbers and murderers; and a tax was laid on the inhabitants for keeping and maintaining a certain number of these animals. But as the arm of Page 260 justice is now extended over every part of the country, and as there are now no secret recesses where villany can be concealed, their services in this respect are become no longer necessary.

"Some few of these dogs, however, are yet kept in the northern parts of the kingdom, and in the lodges of the royal forests, where they are used in pursuit of deer that have been previously wounded. They are also sometimes employed in discovering deer-stealers, whom they infallibly trace by the blood that issues from the wounds of their victims.

"A very extraordinary instance of this occurred in the New Forest, in the year 1810, and was related to me by the Right Hon. G. H. Rose. A person, in getting over a stile into a field near the Forest, remarked that there was blood upon it. Immediately afterwards he recollected that some deer had been killed, and several sheep stolen in the neighbourhood; and that this might possibly be the blood of one that had been killed in the preceding night. The man went to the nearest lodge to give information; but the keeper being from home, he was under the necessity of going to Rhinefield Lodge, which was at a considerable distance. Toomer, the under-keeper, went with him to the place, accompanied by a bloodhound. The dog, when brought to the spot, was laid on the scent; and after following for about a mile the track which the depredator had taken, he came at last to a heap of furze fagots belonging to the family of a cottager. The woman of the Page 261 house attempted to drive the dog away, but was prevented; and on the fagots being removed a hole was discovered in the ground, which contained the body of a sheep that had recently been killed, and also a considerable quantity of salted meat. The circumstance which renders this account the more remarkable is, that the dog was not brought to the scent until more than sixteen hours had elapsed after the man had carried away the sheep."

An old writer—the author of "The History of the Buccaneers"—though full of prejudice against the Indians, thus describes some of the atrocities practised by the Spaniards:—

"The Spaniards having possessed themselves of these isles (South America), found them peopled with Indians, a barbarous people, sensual and brutish, hating all labour, and only inclined to killing and making war against their neighbours; not out of ambition, but only because they agreed not with themselves in some common terms of language; and perceiving that the dominion of the Spaniards laid great restrictions upon their lazy and brutish customs, they conceived an irreconcileable hatred against them, but especially because they saw them take possession of their kingdoms and dominions. Hereupon they made against them all the resistance they could, everywhere opposing their designs to the utmost; and the Spaniards, finding themselves cruelly hated by the Indians, and nowhere secure from Page 262 their treacheries, resolved to extirpate and ruin them, since they could neither tame them by civility nor conquer them by the sword. But the Indians, it being their custom to make the woods their chief places of defence, at present made these their refuge whenever they fled from the Spaniards: hereupon those first conquerors of the New World made use of dogs to range and search the intricate thickets of woods and forests for those their implacable and unconquerable enemies; thus they forced them to leave their old refuge and submit to the sword, seeing no milder usage would do it: hereupon they killed some of them, and quartering their bodies, placed them in the highways, that others might take warning from such a punishment. But this severity proved of ill consequence, for instead of frightening them, and reducing them to civility, they conceived such horror of the Spaniards, that they resolved to detest and fly their sight for ever; hence the greatest part died in caves and subterraneous places of woods and mountains, in which places I myself have often seen great numbers of human bones."

It has been already stated, that in the West Indies bloodhounds were employed to hunt the runaway blacks. I had one of these Cuban bloodhounds given to me a few years ago, and finding him somewhat more ferocious than I liked, I made a present of him to a keeper in the neighbourhood. He was put into Page 263 a kennel with other dogs, and soon killed some of them. Keepers, however, in going their rounds at night, are frequently accompanied by bloodhounds, and poachers are said to have a great dread of them.

Page 264

THE TERRIER.

"Little favourite! rest thee here,
With the tribute of a tear!
*   *   *   *

Thou hast fondled at my feet,
Greeted those I lov'd to greet;
When in sorrow or in pain,
On my bosom thou hast lain.
I have seen thy little eye
Full as if with sympathy."

There are so many varieties of terriers, and so many celebrated breeds of these dogs, that it would be a difficult task to give a separate account of each. Some Page 265 have a cross of the bull-dog; and these, perhaps, are unequalled for courage and strength of jaw. In the latter quality they are superior to the bull-dog. Then there is the pepper-and-mustard breed, the Isle of Sky, the rough and smooth English terrier, and a peculiar breed, of which my own sensible little Judy, now reposing at my feet, is one, besides some others.

Perhaps there is no breed of dogs which attach themselves so strongly to man as the terrier. They are his companions in his walks, and their activity and high spirit enable them to keep up with a horse through a long day's journey. Their fidelity to their master is unbounded, and their affection for him unconquerable. When he is ill they will repose for hours by the side of his bed, as still as a mother watching over a sick and slumbering child; and when he is well they will frisk around him, as if their pleasure was renewed with his returning health. How well do I remember this to have been the case with my faithful old dog Trim! Nothing would induce him to make the slightest noise till I called him on my bed, when I awoke in the morning. Night or day, he never left me for many years; and when at last I was obliged to take a journey without him, his life fell a sacrifice to his affection for me. Alas, poor Trim!

This breed of dogs, the true English terrier, shows an invincible ardour in all that he is required to do, as well as persevering fortitude. In drawing badgers and foxes from their holes, the severe bites of these animals Page 266 only seem to animate them to greater exertions; and they have been known to suffer themselves to be killed by the former sooner than give over the unequal contest.

The vignette at the end of this notice represents a favourite wire-haired terrier of mine, called Peter, well known for many years at Hampton Court. He had wonderful courage and perseverance, and was the best dog to hunt rabbits in thick hedge-rows I ever met with. He was also a capital water-dog; and he was frequently enticed by some of the officers quartered at Hampton Court to accompany them to the neighbouring lock of the river Thames, in which an unfortunate duck was to be hunted. I was assured that on these occasions Peter distinguished himself greatly, diving after the duck whenever it dived, and beating all the other dogs by his energy and perseverance. Peter was a general favourite, and perhaps this was partly owing to his being a great pickle. He was always getting into scrapes. Twice he broke either his shoulder-bone or his leg by scrambling up a ladder. He was several times nearly killed by large dogs, of which he was never known to show the slightest fear; and with those of about his own size he would fight till he died. He has killed sixty rats in a barn in about as many minutes; and he was an inveterate foe to cats. I remember once taking him with me on a rabbit-ferreting excursion. Before the ferrets were put in the holes, I made Peter quite aware that he Page 267 was not to touch them; and he was so sensible a dog that there was no difficulty in doing this, although it was the first time he had seen a ferret. If a rabbit bolted from the hole he was watching, he killed it in an instant; but when the ferret made its appearance, Peter retreated a step or two, showing his teeth a little as if he longed to attack it. Towards the end of the day I had gone to a little distance, leaving Peter watching a hole. Presently I heard a squeak, and on turning round I saw the ferret dead, and Peter standing over it, looking exceedingly ashamed at what he had done, and perfectly conscious that he had disobeyed orders. The temptation, however, was too great for him to resist. Peter at last got into bad company, for he suffered himself to be enticed by the ostlers and others into the taps at Hampton Court, and they indulged him in his fondness for killing vermin and cats. He was a dog of extraordinary sense. I once gave him some milk and water at my breakfast, which was too hot. He afterwards was in the habit of testing the heat by dipping one of his paws into the basin, preferring rather to scald his foot than to run the risk of burning his tongue. He had other peculiarities. When I mounted my horse and wanted him to follow me, he would come a little distance, and then all at once pretend to be lame. The more I called the lamer he became. He was, in fact, aware of my long rides, and was too lazy to follow me. He played this trick very frequently. If I called him while I had my Page 268 snuff-box in my hand, he would come to me, pretending to sneeze the whole of the time. I have said so much about Peter, because he was a good specimen of one of the small breed of terriers.

Terriers, more than any other breed of dogs, live so much in our rooms, and are so generally our companions during our walks and rides, that they naturally imbibe a great degree of sensibility of the least look or word of their master. This very sensibility makes them extremely jealous of any preference or attention shown by their master to another dog. I had an old terrier who never could bear to see me do this. He showed it not only by his countenance in a remarkable way, but would fall upon any dog he saw me caress. Mons. Blaze gives an instance of a dog having killed a young child, who had been in the habit of fondling a dog belonging to the same owner, and showing fear and dislike of him. Another dog was so strongly attached to his master that he was miserable when he was absent. When the gentleman married, the dog seemed to feel a diminution of affection towards him, and showed great uneasiness. Finding, however, that his new mistress grew fond of him, he became perfectly happy. Somewhat more than a year after this they had a child. There was now a decided inquietude about the dog, and it was impossible to avoid noticing that he felt himself miserable. The attention paid to the child increased his wretchedness; he loathed his food, and nothing could content him, though he was Page 269 treated on this account with the utmost tenderness. At last he hid himself in the coal-cellar, and every means were used to induce him to return, but all in vain. He was deaf to entreaty, rejected all kindness, refused to eat, and continued firm in his resolution, till exhausted nature yielded to death.

I have seen so much of the sensitiveness and jealousy of dogs, owing to their unbounded affection for their masters, that I cannot doubt the truth of this anecdote, which was related by Mr. Dibdin. A lady had a favourite terrier, whose jealousy of any attentions shown to her by strangers was so great, that in her walks he guarded her with the utmost care, and would not suffer any one to touch her. The following anecdote will prove the unchanging affection of these dogs. It was communicated to me by the best and most amiable man I have ever met with, either in public or private life.

He had a small terrier, which was much attached to him. On leaving this country for America, he placed the dog under the care of his sister, who resided in London. The dog at first was inconsolable, and could scarcely be persuaded to eat anything. At the end of three years his owner returned, and upon knocking at the door of his sister's house, the dog recognised the well-known knock, ran down-stairs with the utmost eagerness, fondled his master with the greatest affection; and when he was in the sitting-room, the faithful animal jumped upon the piano-forte, that Page 270 he might get as near to him as possible. The dog's attachment remained to the last moment of his life. He was taken ill, and was placed in his master's dressing-room on one of his cloaks. When he could scarcely move, his kind protector met him endeavouring to crawl to him up the stairs. He took the dog in his arms, placed him on his cloak, when the dog gave him a look of affection which could not be mistaken, and immediately died. There can, I think, be no doubt but that this affectionate animal, in his endeavour to get up the steps to his master, was influenced by sensations of love and gratitude, which death alone could extinguish, and which the approach of death prompted him to show. How charming are these instances of the affection of dogs to a kind master! and how forcibly may we draw forth the strongest testimonials of love from them, by treating them as they deserve to be treated! Few people sufficiently appreciate the attachment, fidelity, and sagacity of these too-often persecuted animals, or are aware how much they suffer from unkindness or harsh treatment.

Every one is acquainted with the pretty picture Sir Walter Scott has drawn of the affectionate terrier, which was the companion of his hero in "Guy Mannering." We see the faithful Wasp "scampering at large in a thousand wheels round the heath, and come back to jump up to his master, and assure him that he participated in the pleasures of the journey." We see him during the fight with the robbers, "annoying Page 271 their heels, and repeatedly effecting a moment's diversion in his master's favour, and pursuing them when they ran away." We hear the jolly farmer exclaim—"De'il, but your dog's weel entered wi' the vermin;" and when he goes to see his friend in prison, and brings Wasp with him, we see the joy of the latter, and hear the remark elicited by it—"Whisht, Wasp—man! Wow, but he's glad to see you, poor thing." The whole race of pepper-and-mustard are brought before us—that breed which are held in such high estimation, not only as vermin-killers, but for their intelligence and fidelity, and other companionable qualities.

I could not deny myself the pleasure of introducing this account of the terrier, as it describes so well their courage, fidelity, and attachment. "Wasp," we are told, at the close of an eventful day, "crouched himself on the coverlet at his master's feet, having first licked his master's hand to ask leave." This is part of the natural language of the dog, and how expressive it is! They speak by their eyes, their tail, and by various gestures, and it is almost impossible to misunderstand their meaning. There is a well-known anecdote of two terriers who were in the habit of going out together to hunt rabbits. One of them got so far into a hole that he could not extricate himself. His companion returned to the house, and by his importunity and significant gestures induced his master to follow him. He led him to the hole, made him Page 272 understand what was the matter, and his associate was at last dug out.

The following affords another proof of the sagacity of these dogs:—

A respectable farmer, residing in a village near Gosport, had a terrier dog who was his constant companion. His business frequently led him across the water to Portsmouth, to which place the dog regularly attended him. The farmer had a son-in-law, a bookseller at Portsmouth, to whose house he frequently went, taking the dog with him. One day, the animal having lost his master in Portsmouth, after searching for him at his usual haunts, went to the bookseller, and by various gesticulations gave him to understand that he had lost his master; his supplications were not in vain, for the bookseller, who understood his language, immediately called his boy, gave him a penny, and ordered him to go directly to the beach, and give the ferryman the money for his passage to the opposite shore. The dog, who seemed to understand the whole proceeding, was much pleased, and jumped directly into the boat, and when landed at Gosport, immediately ran home. He always afterwards went to the bookseller, if he had lost his master at Portsmouth, feeling sure that his boat-hire would be paid, and which was always done.

The same dog, when he was wet or dirty, would go into the barn till he was clean and dry, and then scratch at the parlour-door for admittance.

Page 273 The Rev. Leonard Jenyns, in his "Observations in Natural History," records the following.—

"A lady, [Q] living in the neighbourhood of my own village, had some years back a favourite Scotch terrier, which always accompanied her in her rides, and was also in the habit of following the carriage to church every Sunday morning. One summer day the lady and her family were from home several weeks, the dog being left behind. The latter, however, continued to come to church by itself for several Sundays in succession, galloping off from the house at the accustomed hour, so as to arrive at the time of service commencing. After waiting in the churchyard a short time, it was seen to return home quiet and dispirited. The distance from the house to the church is three miles, and beyond that at which the ringing of the bells could be ordinarily heard. This was probably an instance of the force of habit, assisted by some association of recollections connected with the movements of the household on that particular day of the week."

An old house being under repair, the bells on the ground-floor were taken down. The mistress of the house had an old favourite terrier, and when she wanted her servants, sent the dog to ring the bell in her dressing-room, having previously attached a bit of wood to the bell-rope. When the dog pulled at the rope, he listened, and if the bell did not ring, he pulled Page 274 till he heard it, and then returned to the room he had left. If a piece of paper were put into his mouth, with a message written on it, he would carry it to the person he was told to go to, and waited to bring back the answer.

Mr. Laing, who was steward to General Sharp, of Houston, near Uphall, had a terrier dog which gave many proofs of his sagacity. Upon one occasion his wife lent a white petticoat to a neighbour in which to attend a christening; the dog observed his mistress make the loan, followed the woman home who borrowed the article, never quitted her, but accompanied her to the christening, and leaped several times on her knee: nor did he lose sight of her till the piece of dress was at last fairly restored to Mrs. Laing. During the time this person was at the christening she was much afraid the dog would attempt to tear the petticoat off her, as she well knew the object of his attendance.

One of the most extraordinary terriers I ever met with belonged to a man named T— —y, well known for many years in the neighbourhood of Hampton Court. The father of this man had been in a respectable way of life, but his son wanted steadiness of character, and, indeed, good conduct, and had it not been for the kindness of his late Majesty, King William the Fourth, he would have been reduced to poverty long before he was. T— —y, through the interest of the king, then Duke of Clarence, was tried in several Page 275 situations, but failed in them all. At last he was made a postman, but was found drunk one evening with all his letters scattered about him, and, of course, lost his situation. He then took up the employment of rat-catcher, for which, perhaps, he was better qualified than any other. His stock-in-trade consisted of some ferrets and an old terrier dog, and a more extraordinary dog was seldom seen. He was rough, rather strongly made, and of a sort of cinnamon colour, having only one eye; his appearance being in direct contrast to what Bewick designates the *genteel* terrier. The other eye had a fluid constantly exuding from it, which made a sort of furrow down the side of his cheek. He always kept close to the heels of his master, hanging down his head, and appearing the personification of misery and wretchedness. He was, however, a wonderful vermin-killer, and wherever his master placed him, there he remained, waiting with the utmost patience and resignation till an unfortunate rat bolted from the hole, which he instantly killed in a most philosophical manner. The poor dog had to undergo the vicissitudes of hard fare, amounting almost to starvation, of cold, rain, and other evils, but still he was always to be seen at his master's feet, and his fidelity to him was unshaken. No notice, no kind word, seemed to have any effect upon him if offered by a stranger, but he obeyed and understood the slightest signal from his owner. This man was an habitual drunkard, at least whenever he could procure the means of becoming one. It was Page 276 a cold, frosty night in November, when T— —y was returning from a favourite alehouse, along one of the Thames Ditton lanes, some of which, owing to the flatness of the country, have deep ditches by their sides. Into one of these the unfortunate man staggered in a fit of brutal intoxication, and was drowned. When the body was discovered the next morning, the dog was seen using his best endeavours to drag it out of the ditch. He had probably been employed all night in this attempt, and in his efforts had torn the coat from the shoulders of his master. It should be mentioned that this faithful animal had saved his master's life on two former occasions, when he was in nearly similar circumstances.

It may interest some of the readers of this little story to be informed, that a few years before the event which has been related took place, the unhappy man's wife died, leaving four very young children. She was a most industrious woman, of excellent character, and her great misery on her death-bed was the reflection that these children—two boys and two girls—would be left to the care of her drunken husband. She was comforted, however, in her dying moments, by one whose heart and hand have always been ready to relieve the distressed, with the assurance that her children should be taken care of. So when the excellent Queen Adelaide heard of the circumstance, she immediately sent for the four children, placed them under the charge of a proper person, educated and maintained them, placed them in respectable situations in life, and Page 277 continued to be their friend till her death. This is one of numerous instances which could be related by the author of her Majesty's silent, but unbounded benevolence.

It is time, however, to resume my anecdotes of terriers.

A gentleman of my acquaintance had a favourite dog of this description, which generally slept in his bed-room. My friend was in the habit of reading in bed. On calling upon him one morning, he took me into his bed-room, and showed me his bed-curtains much burnt, and one of his sheets. The night before he had been reading the newspaper in bed, with a candle near him, and had gone to sleep. The newspaper had fallen on the candle, and thus set fire to the curtain. He was awoke by his dog scratching him violently with his fore-feet, and was thus in time to call for assistance, and save the house from being burnt down, and also probably to save his own life.

Another of my acquaintances has a very small pet terrier, a capital rat-killer, who always evinces great antipathy to those animals. She lately produced three puppies, two of which were drowned. After hunting for them in every direction, she returned to her litter, where she was found the next morning not only suckling her own whelp, but a young rat; and thus she continued to do till it reached maturity. The morning on which her puppies were drowned there had been a battue of rats, some of which were wounded and escaped. Page 278 One of these latter was the young rat in question. This, no doubt, was taken possession of for the purpose of relieving her of her superabundant milk.

A gentleman who had befriended an ill-used terrier acquired such an influence over the grateful dog, that he was obedient to the least look or sign of his master, and attached himself to him and his children in a most extraordinary manner. One of the children having behaved ill, his father attempted to put the boy out of the room, who made some resistance. The dog seeing the bustle, supposed his master was going to beat the boy, and therefore tried to pull him away by the skirts of his coat, thus showing his affection and sagacity at the same time.

Captain Brown relates the following: —

Sir Patrick Walker writes me: — "Pincer, in appearance, is of the English terrier breed, but in manner indicates a good deal of the Scotch colley, or shepherd's dog. He has a remarkably good nose, is a keen destroyer of vermin, and is in the habit of coming to the house for assistance ever since the following occurrence: — He came into the parlour one evening when some friends were with us, and looking in my face, by many expressive gestures, evinced great anxiety that I should follow him. Upon speaking to him, he leaped, and his whine got to a more determined bark, and pulled me by the collar or sleeve of the coat, until I was induced to follow him; and when I got up, he began leaping and gambolling Page 279 before me, and led the way to an outhouse, to a large chest filled with pieces of old wood, and which he continued by the same means to solicit to be moved. This was done, and he took out a large rat, killed it, and returned to the parlour quite composed and satisfied.

"Similar occurrences have frequently taken place since, with this addition, that as I sometimes called the servant, he often leaves me and runs in the same manner to get his assistance, as soon as he finds me quitting the room to follow him. In no instance has Pincer ever been wrong, his scent is so very good. Once, when he had got assistance, he directed our attention to some loose wood in the yard; and when part of it was removed, he suddenly manifested disappointment, and that the object of pursuit was gone. His manner and look seemed more than instinct, and at once told his story. After a little pause, and some anxious looks, he dashed up a ladder that rested against a low out-house, and took a large rat out of the spout, whither it had apparently escaped whilst Pincer came for assistance."

Terriers appear to have a strong instinctive faculty of finding their way back to their homes, when removed from them to long distances, and even when they have seas to cross. There are instances of their having done this from France, Ireland, and even Germany. Their powers of endurance, therefore, must be very great, and their energies as well as affections equally strong. Page 280 They have also an invincible perseverance in all they do, to which every fox-hunter will bear his testimony. In my youth, when following the hounds, I have been delighted in witnessing the energy of a brace of terriers, who were sure to make their appearance at the slightest check, running with an ardour quite extraordinary, and incessant in their exertions to be with the busiest of the pack in their endeavours to find. If the fox takes to earth, the little brave terrier eagerly follows, and shows by his baying whether the fox lays deep or not, so that those who are employed in digging it out can act accordingly. In rabbit-shooting in thick furze or breaks, the terrier, as I have often witnessed, will take covert with the eagerness and impetuosity of a foxhound. On one of these occasions I saw an enormous wild cat started, which a small terrier pursued and never quitted, notwithstanding the unequal contest, till it was shot by a keeper. As vermin-killers, they are superior to all other dogs. The celebrated terrier Billy was known to have killed one hundred rats in seven minutes.

Nor are their affections less strong than their courage. A gentleman in the neighbourhood of Bath had a terrier which produced a litter of four puppies. He ordered one of them to be drowned, which was done by throwing it into a pail of water, in which it was kept down by a mop till it appeared to be dead. It was then thrown into a dust-hole, and covered with ashes. Two mornings afterwards, the servant disco Page 281 vered that the bitch had still four puppies, and amongst them was the one which it was supposed had been drowned. It was conjectured that in the course of a short time the terrier had, unobserved, raked her whelp from the ashes, and had restored it to life.

An excellent clergyman, residing close to Brighton, gave me the following curious anecdote of a dog which his son, the late greatly-lamented Major R— — brought to England with him from Spain. This dog was a sort of Spanish terrier, and his disposition and habits were very peculiar indeed, unlike those of any dog I ever heard of. One day a teacher of music was going to one of her pupils, and as she was passing at some little distance from the house of the owner of this dog, had her attention attracted to him. He first looked at her very significantly, pulled her by the gown the contrary way to which she was going, and evidently wanted her to follow him. Partly instigated by curiosity, but chiefly because he held her gown tight in his mouth, she suffered herself to be led some distance, when the dog brought her into a field in which some houses were in the course of being built. She then became alarmed, and seeing two or three labourers, she asked them to drive away the dog. Finding, however, that he would not quit his hold, they advised her to see where the dog would lead her, promising to accompany and protect her. Thus assured, she allowed him to lead her where he pleased. Page 282 The dog brought her to the houses which were being built. On arriving at them, it was found that the area had been dug out, and a strong plank placed across it, one end resting on a heap of earth. At this end the dog began to scratch eagerly; and on the plank being lifted up, a large beef bone was discovered, which the dog seized in his mouth, and trotted away with it perfectly satisfied. My informant said that he had taken some pains to ascertain the accuracy of this anecdote from the young lady herself, and that I might depend on its truth.

A somewhat similar occurrence took place in my own neighbourhood, very recently. A lady, going to make a morning's call, passed the gateway of a house, when her gown was seized by a dog, who pulled her the contrary way to which she was going. She at last disengaged herself, and made her call. On coming out, the dog was waiting for her, and again took her gown in his mouth, and led her to the gateway she had previously passed. Here he stopped, and as the dog held a tight hold, she rang the bell; and on a servant opening the gate the animal, perfectly satisfied, trotted in, when she found that he belonged to the house, but had been shut out.

It may be also mentioned as an instance of courage and fidelity in a terrier, that as a gentleman was returning home, a man armed with a large stick seized him by the breast, and striking him a violent blow on the head, desired him instantly to deliver his watch Page 283 and money. As he was preparing to repeat the blow, the terrier sprung at him, and seized him by the throat. His master, at the same time, giving the man a violent blow, he fell backwards and dropped his stick. The gentleman took it up, and ran off, followed by his dog, but not before the animal had torn off and carried away in his mouth a portion of the man's waistcoat.

The following fact will serve to prove that dogs are capable of gratitude in no ordinary degree: —

A surgeon at Dover, seeing a terrier in the street which had received some injury, took it home; and having cured it in a couple of days, let it go. For many weeks the grateful animal used to pay him a daily visit of a few minutes, and after a vehement wagging of his tail, scampered off again to his own home.

A neighbour of mine has a terrier which has shown many odd peculiarities in his habits. He has contracted a great friendship for a white cat, and evinced his affection for it the other day in a curious manner. The dog was observed to scratch a large deep hole in the garden. When he had finished it he sought out the cat, dragged her by the neck to the hole, endeavoured to place her in it, and to cover her with the soil. The cat, not liking this proceeding, at last made her escape.

While two terriers were hunting together in a wood, one was caught by the leg in a trap set for foxes. His companion finding that he could not extricate the Page 284 other, ran to the house of his owner, and by his significant gesticulations induced him to follow; and by this means he was extricated.

Mr. Morritt, well known to the readers of the Life of Sir Walter Scott, as his intimate and confidential friend, had two terriers of the pepper-and-mustard breed, or rather, as we prefer him to any other character Sir Walter Scott has delighted us with, the Dandy Dinmont breed. These dogs (for we avoid the feminine appellation when we can) were strongly attached to their excellent master, and he to them. They were mother and daughter, and each produced a litter of puppies about the same time. Mr. Morritt was seriously ill at this period, and confined to his bed. Fond as these dogs were of their puppies, they had an equal affection for their master, and in order to prove to him that such was the case, they adopted the following expedient. They conveyed their two litters of puppies to one place, and while one of the mothers remained to suckle and take care of them, the other went into Mr. Morritt's bedroom and continued there from morning until the evening. When the evening arrived, she went and relieved the other dog, who then came into the bedroom, and remained quietly all night by the side of the bed, and this they continued to do day after day in succession.

This charming anecdote was communicated to me from a quarter which cannot leave a doubt of its authenticity, and affords an affecting proof of gratitude Page 285 and love in animals towards those who have treated them with kindness, and made them their friends. Such an anecdote as this should be sufficient to preserve dogs from much of the ill-treatment they meet with.

I knew a very clever terrier belonging to a friend of mine. His name was Snap. Now Snap one fine, hot, summer's day, accompanied his master, who was on horseback, on his way from London to the neighbourhood of Windsor. The road was very dusty, and, as I have said, the weather hot, and Snap was very thirsty. No water was met with until Hounslow had been passed. At last a woman crossed the road with a bucket of water, which she had drawn from a neighbouring pump. On arriving at her cottage she placed it outside her door, and left it there. Snap saw it and lapped up some of the water with evident satisfaction, his master waiting for him.

When he had finished his lapping, instead of following, he deliberately inserted his hind-quarters into the bucket—took a good cooling bath—shook himself in the bucket—jumped out—gave himself another shake, and then followed his master. If Snap was lost in London, he would go to every house usually frequented by his master; and if he then could not find him, would return home. Snap, in fact, was an extraordinary dog.

One night, a gentleman, between fifty and sixty years of age, went into a house of a particular de Page 286 scription near the Admiralty. He had not been long there when he died suddenly. He had with him a small dog of the terrier kind, which immediately left the room. There was nothing found on the gentleman's person to lead to a discovery of his name or residence. About twelve o'clock, however, on the following night, three interesting young ladies, of very genteel appearance, between the ages of sixteen and twenty, arrived at the house in which the gentleman died, accompanied by the dog. They came in a chaise from Richmond. It appears that the dog, immediately after the decease of his master, ran off to Richmond, where he usually resided. As soon as the door was opened he rushed into the apartment of the young ladies, who were in the act of dressing themselves. He began to solicit their attention by whines and cries, and his eyes turned to the door, as if to invite them to follow him. Failing in this, he became more earnest, seized their clothes, and pulled them towards the door with so much violence, that one of their gowns was torn. This excited great alarm; and from the intelligence shown by the animal, it was resolved by the young ladies to resign themselves to the dog, which continued to entice them away. A chaise was accordingly ordered, and they immediately took their seats in it. The dog led the way, with its head almost constantly turned back, and his eyes fixed upon the carriage, until he led them to the house near the Admi Page 287 ralty, where his master had died. There they alighted; but how great was their grief, horror, and surprise, to find their father dead in such a situation!

The deceased proved to be Mr. ——, an inhabitant of Lewisham, in Kent, where he possessed a farm of considerable extent, and followed the business of an auctioneer, and was greatly respected in his neighbourhood. That night he dropped down in the house alluded to, when the people, supposing him dead, immediately gave the alarm, and the body was conveyed to the Lord Cochrane hotel, within a few doors, in Spring Gardens. Here it was discovered that the spark of life was not totally extinguished. He was carried up-stairs and put to bed, and medical assistance was called in; but in vain,—in a few minutes he was a corpse. As the people of the house were carrying him up-stairs, a sum of 1100*l.* fell from his pocket in bank-notes, tied up in a bundle, and marked on the outside, "To be paid into Snow's,"—a circumstance sufficient in itself to show that he had not been dishonestly treated by the female who accompanied him into the house from which he was brought, or any other person belonging to it. The interesting little dog, after his return, remained at his post, the faithful guardian of his beloved master's remains. He lay on the foot of the bed, with his eyes constantly fixed on the body, with an eager, anxious, melancholy expression.

The place was crowded with people, led by curiosity to this interesting scene. The dog never appeared Page 288 to take any notice of these strange visitors, and no rude hand attempted to interrupt the little mourner in his melancholy office. The verdict of the coroner's inquest was,—"Died by the visitation of God."

Another of the same breed of dogs evinced much sagacity on the following occasion:—

His master occupied furnished lodgings near the Inns of Court in London. In the hurry of removing from them, neither he nor his servants thought of the dog, who was not in the way when they quitted the house. When the dog returned to it, finding his master gone, he trotted off to Kensington, where an intimate friend of his master resided, and very quietly and patiently made himself at home in the house. As he was well known, he was fed and taken care of, and at the end of three days his master called, and he then gladly went away with him.

In this instance it is, I think, evident, that the dog possessed a sort of reasoning faculty, which induced him to suppose that the best chance he had of finding his master was by going to a place to which he had formerly accompanied him; and he was correct in his calculation.

This faculty was again exercised in the following manner:—

A gentleman residing in the Tower of London had a terrier which he one day lost, about seven miles from town. The dog attached himself to a soldier, and notwithstanding the man went to town in an omnibus, the Page 289 dog followed the vehicle. When the soldier alighted from it, he went to the barracks in St. James's Park, the dog continuing close behind him. On examining the collar, the name and residence of the owner of the dog were found on it. The soldier therefore brought him to the Tower, and gave the above particulars. From this account it may be supposed that the dog, having been familiar with the sight of Guardsmen at the Tower, had followed one of them in hopes that he belonged to that place, and therefore would conduct him to it.

I am not aware that any writer upon dogs has noticed one of their peculiarities, that of curiosity. Let me give a curious and well-authenticated instance of this property, which was communicated to me by the owner of the dog. This animal was a Scotch terrier, named Snob, and certainly a more singular dog has seldom been met with. His master was commander of the fleet on the South American station, and Snob embarked with him. He soon began to give proofs of his extraordinary curiosity, for he liked to see everything that was going forward in the ship. Snob, in fact, was a sort of Paul Pry. He watched everything that was to be done. One night the sailors were kept up aloft for some hours doing something to the sails; Snob remained on the deck the whole time, looking very wise, and watching the sailors with one paw lifted up. He would at other times wander between the decks, looking at everything going forward; and when he had been shut Page 290 in the cabin he has frequently been observed standing on his hind legs looking through the keyhole of the door, in order to watch the proceedings which were carried on. I have a great respect for Snob, who is still alive, and I have no doubt his curiosity is as great as ever.

A curious instance of ferocity and affection in a terrier bitch is recorded by Mr. Daniel:—After a very severe burst of upwards of an hour, a fox was, by Mr. Daniel's hounds, run to earth, at Heney Dovehouse, near Sudbury, in Suffolk. The terriers were lost; but as the fox went to ground in view of the headmost hounds, and it was the concluding day of the season, it was resolved to dig him out, and two men from Sudbury brought a couple of terriers for that purpose. After considerable labour, the hunted fox was got, and given to the hounds; whilst they were breaking him, one of the terriers slipped back into the earth, and again laid. After more digging, a bitch-fox was taken out, and the terrier killed two cubs in the earth; three others were saved from her fury, and which were begged by the owner of the bitch, who said he should make her suckle them. This was laughed at as impossible; however, the man was positive, and the cubs were given to him. The bitch-fox was carried away, and turned into an earth in another county. The terrier had behaved so well at earth, that she was some days afterwards bought, with the cubs she had fostered, by Mr. Daniel. The bitch continued regularly to suckle, and reared them until able to shift for themselves. What adds to this singu Page 291 larity is, that the terrier's whelp was nearly five weeks old, and the cubs could just see, when this exchange of progeny was made.

The following is a proof not only of the kind disposition, but the sense of a terrier.

A gentleman, from whom I received the anecdote, was walking one day along a road in Lancashire, when he was *accosted*, if the term may be used, by a terrier dog. The animal's gesticulations were at first so strange and unusual, that he felt inclined to get out of its way. The dog, however, at last, by various significant signs and expressive looks, made his meaning known, and the gentleman, to the dog's great delight, turned and followed him for a few hundred yards. He was led to the banks of a canal, which he had not before seen, and there he discovered a small dog struggling in the water for his life, and nearly exhausted by his efforts to save himself from drowning. The sides of the canal were bricked, with a low parapet wall rather higher than the bank. The gentleman, by stooping down, with some difficulty got hold of the dog and drew him out, his companion all the time watching the proceedings. It cannot be doubted, but that in this instance the terrier made use of the only means in his power to save the other dog, and this in a way which showed a power of reasoning equally strong with that of a human being, under a similar circumstance.

I may here mention another instance of a terrier finding his way back to his former home.

Page 292 A gentleman residing near York went to London, and on his return brought with him a young terrier dog, which had never been out of London. He brought him to York in one of the coaches, and thence conveyed him to his residence. Impatient of separation from his former master, he took the first opportunity of escaping from the stable in which he had been confined, and was seen running on the turnpike road towards York by the boy who had him in charge, and who followed him for some distance. A few days afterwards, the gentleman who had lost the dog received a letter front London, acquainting him that the dog was found lying at the door of his lodgings, his feet quite sore, and in a most emaciated condition.

A few years ago, a blind terrier dog was brought from Cashiobury Park, near Watford, to Windsor. On arriving at the latter place he became very restless, and took the first opportunity of making his escape, and, blind as he was, made his way back to Cashiobury Park, his native place.

A correspondent informs me, that whilst he was taking a walk one summer's evening, he observed two rough-looking men, having a bull-dog with them, annoying a sickly-looking young gentleman, who was accompanied by a terrier. The bull-dog at last seized the latter, and would soon have killed it, had not my correspondent interfered. He was then informed that a few years previous, when his master was in bed, this little terrier came to his bedroom door, and scratched Page 293 and yelled to be admitted. When this had been done, he immediately rushed to a closet-door in the room, at which he barked most furiously. His master, becoming alarmed, fastened the door, and having obtained the assistance of his servants, a notorious thief was discovered in the closet.

Mr. White, of Selborne, relates a pleasing anecdote of affection, which existed between two incongruous animals—a horse and a hen, and which showed a mutual fellowship and kindness for each other. The following anecdote, communicated to me by a clergyman in Devonshire, affords another proof of affection between two animals of opposite natures. I will give it in his own words:—

"Some few months since it was necessary to confine our little terrier bitch, on account of distemper. The prison-door was constructed of open bars; and shortly after the dog was placed in durance, we observed a bantam cock gazing compassionately at the melancholy inmate, who, doubtless, sadly missed its warm rug by the parlour fire. At last the bantam contrived to squeeze through the bars, and a friendship of a most unusual kind commenced. Pylades and Orestes, Nisus and Euryalus, could not have been bound by closer bonds of affection. The bantam scarcely forsook the poor prisoner's cell for its daily food, and when it did the dog became uneasy, whining till her friend returned, and then it was most amusing to watch the actions of the biped and quadruped. Page 294 As the dog became worse, so did the bantam's attentions redouble; and by way of warming the dog, it took its place between the forelegs, and then the little animal settled luxuriously down on the bird, seeming to enjoy the warmth imparted by the feathers. In this position, and nestled closely side by side, did this curious pair pass some weeks, till death put an end to the poor dog and this singular friendship. It must be added for the bantam's honour, that he was most melancholy for some time afterwards."

The same clergyman also communicated to me the following anecdote illustrative of the sagacity of terriers.

He says that "his brother-in-law, who has a house in Woburn Place, and another in the City, had a wire-haired terrier named Bob, of extraordinary sagacity. The dog's knowledge of London and his adventures would form a little history. His master was in the habit, occasionally, of spending a few days at Gravesend, but did not always take his dog with him. Bob, left behind one day against his liking, scampered off to London Bridge, and out of the numerous steamers boarded the Gravesend boat, disembarked at that place, went to the accustomed inn, and not finding his master there, got on board the steamer again and returned to town. He then called at several places usually frequented by his master, and afterwards went home to Woburn Place. He has frequently been stolen, but always returns, sometimes in sad plight, with a broken cord round his neck, and with signs of Page 295 ill-usage; but still he contrives to escape from the dog-stealers."

I once took a favourite terrier with me to a house I had hired in Manchester Street. He had never been in London before. While the carriage was unloading in which the dog had been conveyed, he was missed, and I could hear nothing of him for nearly a fortnight; at the end of that time he found his way back to the house, with a short cord round his neck, which he had evidently gnawed off. How he came to find his way back is not a little to be wondered at. His joy on seeing me again I cannot forget. Poor Peter! when he got old, and my rides became too long for him, he pretended to be lame after accompanying me a short distance, and would then trot back without any appearance of lameness.

The following anecdote proves the kind disposition of a terrier. A kitten, only a few hours old, had been put into a pail of water, in the stable-yard of an inn, for the purpose of drowning it. It had remained there for a minute or two, until it was to all appearance dead, when a terrier bitch, attached to the stables, took the kitten from the water, and carried it off in her mouth. She suckled and watched over it with great care, and it throve well. The dog was at the same time suckling a puppy about ten weeks old, but which did not seem at all displeased with the intruder.

I had once an opportunity of witnessing the sense of a terrier. I was riding on Sunbury Common, where Page 296 many roads diverge, when a terrier ran up, evidently in pursuit of his master. On arriving at one of the three roads, he put his nose to the ground and snuffed along it; he then went to the second, and did the same; but when he came to the third, he ran along it as fast as he could, without once putting down his nose to the ground. This fact has been noticed by others, but I never before witnessed it myself.

At Dunrobin Castle, in Sutherlandshire (then the seat of the Marquis of Stafford now of the Duke of Sutherland), there was to be seen, in May 1820, a terrier bitch nursing a brood of ducklings. She had a litter of whelps a few weeks before, which were taken from her and drowned. The unfortunate mother was quite disconsolate till she perceived the brood of ducklings, which she immediately seized and carried to her lair, where she retained them, following them out and in with the greatest care, and nursing them, after her own fashion, with the most affectionate anxiety. When the ducklings, following their natural instinct, went into the water, their foster-mother exhibited the utmost alarm; and as soon as they returned to land she snatched them up in her mouth, and ran home with them. What adds to the singularity of this circumstance is, that the same animal when deprived of a litter of puppies the year preceding, seized two cock-chickens, which she reared with the like care she bestows upon her present family. When the young cocks began to try their voices, their foster-mother was Page 297 as much annoyed as she now seems to be by the swimming of the ducklings, and never failed to repress their attempts at crowing.

The foreman of a brickmaker, at Erith in Kent, went from home in company with his wife, and left her at the Plough at Northend with his brother, while he proceeded across the fields to inspect some repairs at a cottage. In about an hour after his departure, his dog, a small Scotch terrier, which had accompanied him, returned to the Plough, jumped into the lap of his mistress, pawed her about, and whined piteously. She at first took no particular notice of the animal, but pushed him from her. He then caught hold of her clothes, pulled at them repeatedly, and continued to whine incessantly. He endeavoured, also, in a similar way to attract the attention of the brother. At last all present noticed his importunate anxiety, and the wife then said she was convinced something had happened to her husband. The brother and the wife, with several others, went out and followed the dog, who led them through the darkness of the night, which was very great, to the top of a precipice, nearly fifty feet deep; and standing on the bank, held his head over, and howled in a most distressing manner. They were convinced that the poor man had fallen over; and having gone round to the bottom of the pit, they found him, lying under the spot indicated by the dog, quite dead.

Page 298 The following anecdote is copied from a recent number of "The Field:"—

I well remember, when a boy, at Barton-upon-Humber, a certain "keel" employed in the Yorkshire corn-trade, on board which the captain had a dog, possessed of some traces of terrier blood, smooth-coated, and of a pure white colour, his neck and back adorned with stumpy bristles, which ruffled up at the slightest provocation—altogether he looked a mongrel cur enough, but he was an excellent sailor, for he attended his master on all his trading expeditions, and never deserted his ship. One day, while the keel lay in Barton Haven, the dog was lost, and great was the consternation in consequence. Diligent search was made in the town and neighbourhood, but every effort to discover the missing animal proved unavailing. Month after month passed away, the keel went and came on her accustomed avocations, and poor Keeper was forgotten—considered by his master to be dead. Judge, therefore, the man's surprise when one day steering with difficulty his vessel into Goole Harbour, which was crowded with shipping at the time, his glance suddenly fell upon his faithful and long-lost dog, buffeting the water at a considerable distance from the keel, but making eagerly towards her. By the aid of a piece of tar-rope, which was dangling round the dog's neck, and a friendly boat-hook, he was lifted quite exhausted on to the deck of his master's craft, when it became at Page 299 once apparent that he had long been kept a prisoner, most probably on board a vessel, by some one who had stolen him at Barton. The cause of the poor dog's sudden reappearance was undoubtedly his having heard his master's well-remembered voice; but it is strange he should have been able to distinguish at so great a distance, and when swelling that chorus of hoarse bawling which arises from a hundred husky throats when a Yorkshire keelman is engaged forcing his craft into a crowded harbour; and it is also equally touching, that when roused by the distant sound, the poor beast should have plunged, encumbered as he was with the rope he had just burst asunder, so gallantly into the water—an element he was ill-adapted to move in, and in which his master declared he had never seen him before.

Page 300

THE SPANIEL.

"Though once a puppy, and a fop by name,
Here moulders one whose bones some honour claim;
No sycophant, although of Spanish race,

And though no hound, a martyr to the chase.
Ye pheasants, rabbits, leverets rejoice,
Your haunts no longer echo to his voice;
This record of his fate, exulting view—
He died worn out with vain pursuit of you.
'Yes,' the indignant shade of *Fop* replies,
'And worn with *vain pursuits*, man also dies.'"—Cowper.

Poor Doll! the very name of spaniel reminds me of you. How well do I now see your long pendent ears, your black expressive eyes, your short, well-rounded mouth, your diminutive but strong legs, almost hidden by the long, silky hair from your stomach, and hear Page 301 you sing as you lie on the rug before a good fire in the winter, after a hard day's cock or snipe-shooting, wet and tired with your indefatigable exertions! Yes—strange as it may sound, Doll would sing in her way, as I have stated in a previous page; and such was her sagacity, that in process of time when I said, "Sing, Doll," she gave vent to the sounds, and varied them as I exclaimed, "Louder, louder." All this time she appeared to be fast asleep.—And what a dog she was in thick cover, or in rushy swamps! No day was too long for her, nor could a woodcock or snipe escape her "unerring nose:"—

"Still her unerring nose would wind it—
If above ground was sure to find it."

Monsieur Blaze also tells us, that a gentleman had a dog which he taught to utter a particular musical note, and that the animal made a cry which very much resembled it. He then sounded another note close to the ear of the dog, saying to him, "Too high, or too low," according to the degree of intonation. The animal finished by pretty correctly giving the note which was required.

An account is given in the "Bibliothèque Universelle," of a spaniel, who, if he heard any one play or sing a certain air, "L'âne de notre moulin est mort, la pauvre bête," &c., which is a lamentable ditty, in the minor key, the dog looked very pitifully, then gaped repeatedly, showing increasing signs of impatience and Page 302 uneasiness.

He would then sit upright on his hind-legs, and begin to howl louder and louder till the music stopped. No other air ever affected him, and he never noticed any music till the air in question was played or sung. He then manifested, without exception or variation, the series of actions which have been described.

I knew a dog which howled whenever it was pitied, and another whose ear was so sensitive, that it could never bear to hear me make a moaning noise. I have likewise seen a dog affected by peculiar notes played on a violoncello.

It is only now and then that such dogs as Doll are to be met with, and when they are, they are invaluable, either as sporting dogs or as companions. In the latter capacity Doll was quite delightful. In an early May morning, when she knew that no shooting was going forward, she would frisk around me as I strolled in a meadow, gay with my favourite cowslips, or run before me as I passed along a lane, where primroses were peeping out of its mossy sides, looking back every now and then to see if I was following her. There was the dew still glittering on the flowers, which, from their situation, had not yet felt the influence of the morning sun, reminding me of some favourite lines by my favourite poet, Herrick:—

> "Fall on me like a silent dew,
> Or like those maiden showers,
> Which, by the peep of day, do strew
> *A baptism o'er the flowers.*"

Page 303 How delightful it is to think of these bygone walks, and how pleasant to call to mind these traits of a favourite and faithful animal! The poet Cowper was never more engaging than when he describes his vain attempts to reach the flower of a water-lily, as he was strolling along the banks of a stream attended by his spaniel, and afterwards discovering that the sagacious animal had been in the river and plucked it for him.

Another instance of wonderful sagacity in this breed of dogs may be here noticed.

A gentleman shooting wild fowl one day on a lake in Ireland, was accompanied by a sagacious spaniel. He wounded a wild duck, which swam about the lake, and dived occasionally, followed by the dog. The bird at last got to some distance, and lowered itself in the water, as ducks are known to do when they are wounded and pursued, leaving nothing but his head out of it. The dog swam about for some time in search of his prey, but all scent was lost, and he obeyed his master's call, and returned to the shore. He had no sooner arrived there, however, than he ran with the greatest eagerness to the top of some high ground close to the lake. On arriving there, he was seen looking round in every direction; and having at last perceived the spot where the duck was endeavouring to conceal itself, he again rushed into the water, made directly to the spot he had previously marked, and at last succeeded in securing the wounded bird.

Page 304 A spaniel which had been kindly treated and fed, during the absence of his master, in the kitchen of a neighbour, showed his gratitude not only by greeting the cook when he met her, but on one occasion he laid down at her feet a bird which he had caught, wagged his tail and departed; thus showing that he had not forgotten the favours he had received.

The following old, but interesting anecdote, is taken from Daniel's "Rural Sports:"—

"A few days before the overthrow of Robespierre, a revolutionary tribunal had condemned M. R——, an upright magistrate and a most estimable man, on a pretence of finding him guilty of a conspiracy. His faithful dog, a spaniel, was with him when he was seized, but was not suffered to enter the prison. He took refuge with a neighbour of his master's, and every day at the same hour returned to the door of the prison, but was still refused admittance. He, however, uniformly passed some time there, and his unremitting fidelity won upon the porter, and the dog was allowed to enter. The meeting may be better imagined than described. The gaoler, however, fearful for himself, carried the dog out of the prison; but he returned the next morning, and was regularly admitted on each day afterwards. When the day of sentence arrived, the dog, notwithstanding the guards, penetrated into the hall, where he lay crouched between the legs of his master. Again, at the hour of execution, the faithful dog is there; the knife of the guillotine Page 305 falls—he will not leave the lifeless and headless body. The first night, the next day, and the second night, his absence alarmed his new patron, who, guessing whither he had retired, sought him, and found him stretched upon his master's grave. From this time, for three months, every morning the mourner returned to his protector merely to receive food, and then again retreated to the grave. At length he refused food, his patience seemed exhausted, and with temporary strength, supplied by his long-tried and unexhausted affection, for twenty-four hours he was observed to employ his weakened limbs in digging up the earth that separated him from the being he had served. His powers, however, here gave way; he shrieked in his struggles, and at length ceased to breathe, with his last look turned upon the grave."

The late Rev. Mr. Corsellis, of Wivenhoe, in Essex, had an old gamekeeper who had reared a spaniel, which became his constant companion, day and night. Wherever the keeper appeared Dash was close behind him, and was of infinite use in his master's nocturnal excursions. The game at night was never regarded, although in the day no spaniel could find it in better style, or in a greater quantity. If at night, however, a strange foot entered the coverts, Dash, by a significant whine, informed his master that an enemy was abroad, and thus many poachers have been detected. After many years of friendly companionship the keeper was seized with a disease Page 306 which terminated in death. Whilst the slow but fatal progress of his disorder allowed him to crawl about, Dash, as usual, followed his footsteps; and when nature was nearly exhausted, and he took to his bed, the faithful animal unweariedly attended at the foot of it. When he died the dog would not quit the body, but lay on the bed by its side. It was with difficulty he could be induced to eat any food; and though after the burial he was caressed with all the tenderness which so fond an attachment naturally called forth, he took every opportunity to steal back to the room where his old master died. Here he would remain for hours, and from thence he daily visited his grave. At the end, however, of fourteen days, notwithstanding every kindness and attention shown him, the poor faithful animal died, a victim of grief for the loss of his master.

In recording such an instance of affection, it is impossible not to feel regret that animals capable of so much attachment should ever be subjected to ill-usage. Whenever they are treated with kindness and affection, they are ready to return it four-fold. It is generally ill-treatment which produces ferocity or indifference, and the former must be very great before the love of their master can be conquered.

Mr. Blaine records the following story of a dog which he had found:—

"I one day picked up in the streets an old spaniel bitch, that some boys were worrying, from which her Page 307 natural timidity rendered her incapable of defending herself. Grateful for the protection, she readily followed me home, where she was placed among other dogs, in expectation of finding an owner for her; but which not happening, she spent the remainder of her life (three or four years) in this asylum. Convinced she was safe and well treated, I had few opportunities of particularly noticing her afterwards, and she attached herself principally to the man who fed her. At a future period, when inspecting the sick dogs, I observed her in great pain, occasionally crying out. Supposing her to be affected in her bowels, and having no suspicion she was in pup, I directed some castor-oil to be given her. The next day she was still worse, when I examined her more attentively, and, to my surprise, discovered that a young one obstructed the passage, and which she was totally unable to bring forth. I placed her on a table, and, after some difficulty, succeeded in detaching the puppy from her. The relief she instantly felt produced an effect I shall never forget; she licked my hands, and when put on the ground she did the same to my feet, danced round me, and screamed with gratitude and joy.

"From this time to her death, which did not happen till two years after, she never forgot the benefit she had received; on the contrary, whenever I approached, she was boisterous in evincing her gratitude and regard, and would never let me rest till, by noticing her, I had convinced her that I was sensible of her Page 308 caresses. The difference between her behaviour before this accident and after it was so pointed and striking, that it was impossible to mistake the grateful sense she had ever retained of the kindness which had been shown to her."

Spaniels in cover are merry and cheerful companions, all life and animation. They hunt, they frisk about, watching the movements of their master, and are indefatigable in their exertions to find game for him. Their neat shape, their beautiful coats, their cleanly habits, their insinuating attention, incessant attendance, and faithful obedience, insure for them general favour. It is almost impossible, therefore, not to have the greatest attachment and affection for them, especially as few dogs evince so much sagacity, sincerity, patience, fidelity, and gratitude. From the time they are thrown off in the field, as a proof of the pleasure they feel in being employed, the tail is in perpetual motion, upon the increased vibration of which the experienced sportsman well knows when he is getting nearer to the game. As the dog approaches it, the more energetic he becomes. Tremulous whimpers escape him as a matter of doubt occurs, and he is all eagerness as he hits again on the scent. The Clumber breed of spaniels have long been celebrated for their strength and powers of endurance, their unerring nose, and for hunting mute—a great qualification where game abounds. This breed has been preserved in its purity by the successive Dukes of Newcastle, and may be con Page 309 sidered as an aristocratic apanage to their country seats. Nor should the fine breed of spaniels belonging to the Earl of Albemarle be passed by in silence. They are black and tan, of a large size, with long ears, and very much feathered about the legs. They are excellent retrievers; and those who have seen will not soon forget Sir Edwin Landseer's charming picture of the late Lord Albemarle's celebrated dog Chancellor, and one of his progeny, holding a dead rabbit between them, as if equally eager to bring it to their amiable master. These dogs, like those of the Clumber breed, hunt mute, and seldom range out of shot.

While on the subject of Lord Albemarle's breed of dogs, I may mention an extraordinary fact which I noticed in a former work, and which I witnessed myself. I allude to the circumstance of a favourite dog having died after producing a litter of puppies, which were adopted, suckled, and brought up by a young bitch of the same breed, who never had any whelps of her own, or indeed was in the way of having any. The flow of milk of the foster-mother was quite sufficient for the sustenance of the adopted offspring, and enabled her to support and bring them up with as much care and affection as if they had been her own. Here was an absence of that *notus odor* which enables animals to distinguish their young from those of others, and also of that distension of milk which makes the suckling their young so delightful to them. Indeed it may be observed how beautifully and providentially Page 310 it has been ordered, that the process of suckling their young is as pleasurable to the parent animal as it is essential to the support of the infant progeny. The mammæ of animals become painful when over-distended with milk. Drawing off that fluid removes positive uneasiness and affords positive pleasure. In the present instance, however, nothing of the sort was the case, and therefore we can only look to that kindliness of disposition and intelligence with which many animals are so strongly endowed as the reason of the singular adoption referred to. I am aware that this fact has been doubted, but it is too well known and authenticated to admit of the possibility of any mistake. In this instance it must be allowed that the usually defined bounds of instinct were exceeded. If so, distress at hearing the cries of the helpless young must have acted forcibly on the kindly feelings of a poor brute, and thus induced her to act in the manner I have described.

Spaniels, like other dogs, possess the power of finding their way to their homes from distances of considerable extent, and over ground they have not before traversed.

A lady residing at Richmond (Mrs. Grosvenor) gave the Rev. Leonard Jenyns the following anecdote of a dog and cat. A little Blenheim spaniel of hers once accompanied her to the house of a relative, where it was taken into the kitchen to be fed, when two large favourite cats flew at it several times, and Page 311 scratched it severely. The spaniel was in the habit of following its mistress in her walks in the garden, and by degrees it formed a friendship with a young cat of the gardener's, which it tempted into the house,—first into the hall, and then into the kitchen,—where, on finding one of the large cats, the spaniel and its ally fell on it together, and, without further provocation, beat it well; they then waited for the other, which they served in the same manner, and finally drove both cats from the kitchen. The two friends continued afterwards to eat off the same plate as long as the spaniel remained with her mistress in the house.

A gentleman residing at Worcester had a favourite spaniel, which he brought with him to London inside the coach. After having been in town a day or two he missed the dog, and wrote to acquaint his family at Worcester of his loss. He received an answer informing him that he need not distress himself about "Rose," as she had arrived at her old house at Worcester five days after she had been lost in London, but very thin and out of condition. This same dog was a great favourite, and much domesticated. She formed a friendship with the cat, and when before the fire the latter would lie down in the most familiar manner by the side of the dog. When the dog had puppies, the cat was in the habit of sucking her; and it happened more than once that both had young ones at the same time, when the cat might be seen sucking the bitch, and the kittens taking their nourishment from the cat.

Page 312 A friend of mine, who then resided in South Wales, had a team of spaniels, which he used for woodcock shooting. As he was leaving the country for a considerable length of time, he gave permission to some of his neighbours to take out his spaniels when they wanted them. One of these was a remarkably good dog, but of rather a surly disposition, and had, in consequence, been but little petted or noticed by his master. Notwithstanding this, nothing could induce him either to follow or hunt with those to whom he was lent. In order, therefore, to make him of any use, it was necessary to get his feeder to accompany the shooting party, and the dog would then take to hunt in cover; but if this man returned home, the dog would find it out and be there before him. At the end of nearly six years his master returned into Wales, and near the house discovered his old dog, apparently asleep. Knowing his ferocious disposition, he did not venture to go close to him, but called him by name, which did not appear to excite the animal's attention. No sooner, however, did the dog hear an old exciting *cover-call*, than he jumped up, sprang to his old master, and showed his affection for him in every possible way. When the shooting season came, he proved himself to be as good a dog as ever.

Mons. Blaze says, that a fondness for the chase does not always make a dog forget his fidelity to his master. He was one day shooting wild ducks with a friend near Versailles, when, as soon as the first shot Page 313 was fired, a fine spaniel dog joined and began to caress them. They shot during the whole day, and the dog hunted with the greatest zeal and alacrity. Supposing him to be a stray dog, they began to think of appropriating him to themselves; but as soon as the sport was over, the dog ran away. They afterwards discovered that he belonged to one of the keepers, who was confined to his house by illness. His duty, however, was to shoot ducks on one particular day of the week, when he was accompanied by this spaniel; he lived six miles from the spot, and the dog, knowing the precise day, had come there to enjoy his usual sport, and then returned to his master.

One of the most extraordinary cases on record of a friendship between two most dissimilar animals, a spaniel and a partridge, is narrated by a writer in whom implicit confidence may be placed:— "We were lately (in 1823) visiting in a house, where a very pleasing and singular portrait attracted our observation: it was that of a young lady, represented with a partridge perched upon her shoulder, and a dog with his feet on her arm. We recognised it as a representation of the lady of the house; but were at a loss to account for the odd association of her companions. She observed our surprise, and at once gave the history of the bird and the spaniel. They were both, some years back, domesticated in her family. The dog was an old parlour favourite, who went by the name of Tom; the partridge was more recently introduced from France, and an Page 314 swered to the equally familiar name of Bill. It was rather a dangerous experiment to place them together, for Tom was a lively and spirited creature, very apt to torment the cats, and to bark at any object which roused his instinct. But the experiment was tried; and Bill, being very tame, did not feel much alarm at his natural enemy. They were, of course, shy at first; but this shyness gradually wore off: the bird became less timid, and the dog less bold. The most perfect friendship was at length established between them. When the hour of dinner arrived, the partridge invariably flew on his mistress's shoulder, calling with that shrill note which is so well known to sportsmen; and the spaniel leapt about with equal ardour.

One dish of bread and milk was placed on the floor, out of which the spaniel and bird fed together. After their social meal, the dog would retire to a corner to sleep, while the partridge would nestle between his legs, and never stir till his favourite awoke. Whenever the dog accompanied his mistress out, the bird displayed the utmost disquietude till his return; and once, when the partridge was shut up by accident a whole day, the dog searched about the house, with a mournful cry which indicated the strength of his affection. The friendship of Tom and Bill was at length fatally terminated. The beautiful little dog was stolen; and the bird from that time refused food, and died on the seventh day, a victim to his grief."

A friend of mine has a small spaniel, which very Page 315 recently showed great sagacity. This dog, which is much attached to him, was left under the care of a servant while his master paid a visit of a few weeks in Hampshire. The poor animal was so miserable during his absence, that he was informed of it, and directed the dog to be sent to him in a hamper, which was done. He was overjoyed at the sight of his kind master, and remained perfectly contented at his new abode. When preparations were making for his departure, the day before it took place, the dog was evidently aware of what was going forward, and showed his dread of being again left behind, by keeping as close as possible to the feet of his master during the evening. On getting up very early the next morning, before daylight, he found on opening his door that the apprehensive animal was lying before it, although it was winter, and very cold. At breakfast the dog not only nestled against his feet, but rubbed himself so much against them, that he was at last turned out of the room. On going into his dressing-room, where the dog had been in the habit of sleeping in a warm basket before a good fire, he found him coiled up in his portmanteau, which had been left open nearly packed.

In this instance, the animal's knowledge of what was going forward was very evident, and his fear of being left behind could not be more strongly expressed; thus affording another proof that animals are possessed of a faculty much beyond mere instinct.

A young gentleman lately residing in Edinburgh Page 316 was master of a handsome spaniel bitch, which he had bought from a dealer in dogs. The animal had been educated to steal for the benefit of its protector; but it was some time ere his new master became aware of this irregularity of morals, and he was not a little astonished and teazed by its constantly bringing home articles of which it had feloniously obtained possession. Perceiving, at length, that the animal proceeded systematically in this sort of behaviour, he used to amuse his friends, by causing the spaniel to give proofs of her sagacity in the Spartan art of privately stealing; putting, of course, the shopkeepers where he meant she should exercise her faculty on their guard as to the issue.

The process was curious, and excites some surprise at the pains which must have been bestowed to qualify the animal for these practices. As soon as the master entered the shop, the dog seemed to avoid all appearance of recognizing or acknowledging any connexion with him, but lounged about in an indolent, disengaged, and independent sort of manner, as if she had come into the shop of her own accord. In the course of looking over some wares, his master indicated by a touch on the parcel and a look towards the spaniel, that which he desired she should appropriate, and then left the shop. The dog, whose watchful eye caught the hint in an instant, instead of following his master out of the shop, continued to sit at the door, or lie by the fire, watching the counter, until she observed the Page 317 attention of the people of the shop withdrawn from the prize which she wished to secure. Whenever she saw an opportunity of doing so, as she imagined, unobserved, she never failed to jump upon the counter with her fore feet, possess herself of the gloves, or whatever else had been pointed out to her, and escape from the shop to join her master.

A gentleman lately communicated to me the following fact:—

His avocations frequently took him by the side of St. Bride's Churchyard, in London. Whenever he passed it, in the course of some two or three years, he always saw a spaniel at one particular grave—it was the grave of his master. There, month after month, and year after year, did this faithful animal remain, as if to guard the remains of the being he loved. No cold, however severe, no rain, however violent, no sun, however hot, could drive this affectionate creature from a spot which was so endeared to him. The good-natured sexton of the churchyard, (and the fact is recorded to his honour,) brought food daily to the dog, and then pitying his exposure to the weather, scooped out a hole by the side of the grave, and thatched it over.

The following is from the Percy collection of Anecdotes:—

Two spaniels, mother and son, were self-hunting in Mr. Drake's woods, near Amersham, in Bucks. The gamekeeper shot the mother; the son, frightened, ran Page 318 away for an hour or two, and then returned to look for his mother. Having found her dead body, he laid himself down by her, and was found in that situation the next day by his master, who took him home, together with the body of the mother. Six weeks did this affectionate creature refuse all consolation, and almost all nutriment. He became, at length, universally convulsed, and died of grief.

These two anecdotes would form a pretty picture of fidelity and kindness, and there is one (I need not mention Sir Edwin Landseer) who would do justice to them.

I may here remark, that the dogs of poor people generally show more attachment to their masters than those of the rich. Their fidelity appears greater, and more lasting. Misery would seem to tighten the cord of affection between them. They both suffer the same privations together of hunger, cold, and thirst, but these never shake the affection of a dog for his master. The animal's resignation is perfect, and his love unbounded. How beautifully has Sir Walter Scott described the affection of a dog for his master, who fell down a precipice in a fog near the Helvellyn Mountains, in Cumberland, and was dashed to pieces. It was not till more than three months afterwards that his remains were discovered, when his faithful dog was still guarding them.

"Dark green was the spot 'mid the brown mountain heather,
Where the pilgrim of nature lay stretch'd in decay;
Like the corpse of an outcast abandon'd to weather,
'Till the mountain winds wasted the tenantless clay.
Nor yet quite deserted, though lonely extended,
For faithful in death his mute fav'rite attended,
The much-lov'd remains of his master defended,

And chas'd the hill fox and the raven away."

Page 319 Nor are the preceding anecdotes solitary instances of the affection of dogs for their departed masters. Mr. Youatt, in his work on "Humanity to Brutes," which does him so much credit, has recorded the following fact, very similar to the one already given:—

Opposite to the house of a gentleman, near the churchyard of St. Olave, Southwark, where the receptacles of humanity are in many parts dilapidated, was an aperture just large enough to admit a dog. It led along a kind of sink to a dark cavity, close to which a person had recently been buried. It was inhabited by his dog, who was to be seen occasionally moving into or out of the cavern, which he had taken possession of the day of the funeral. How he obtained any food during the first two or three months no one knew, but he at length attracted the attention of a gentleman who lived opposite, and who ordered his servant regularly to supply the dog with food. He used, after a while, to come occasionally to this house for what was provided for him. He was not sullen, but there was a melancholy expression in his countenance, which, once observed, would never be forgotten. As soon as he had finished his hasty meal, he would gaze for a moment on his benefactor. It was an expressive look, Page 320 but one which could not be misunderstood. It conveyed all the thanks that a broken heart could give. He then entombed himself once more for three or four days, when he crawled out again with his eyes sunk and his coat dishevelled. Two years he remained faithful to the memory of the being he had lost, and then, according to the most authentic account of him, having been missing several days, he was found dead in his retreat.

From a letter written by a gentleman at Dijon in France, to his friend in London, dated August 15, 1764, we have the following account of a murder discovered by a dog:—

"Since my arrival here a man has been broken on the wheel, with no other proof to condemn him than that of a water-spaniel. The circumstances attending it being so very singular and striking, I beg leave to communicate them to you. A farmer, who had been to receive a sum of money, was waylaid, robbed, and murdered, by two villains. The farmer's dog returned with all speed to the house of the person who had paid the money, and expressed such amazing anxiety that he would follow him, pulling him several times by the sleeve and skirt of the coat, that, at length, the gentleman yielded to his importunity. The dog led him to the field, a little from the roadside, where the body lay. From thence the gentleman went to a public-house, in order to alarm the country. The moment he entered, (as the two villains were there Page 321 drinking,) the dog seized the murderer by the throat, and the other made his escape. This man lay in prison three months, during which time they visited him once a-week with the spaniel, and though they made him change his clothes with other prisoners, and always stand in the midst of a crowd, yet did the animal always find him out, and fly at him. On the day of trial, when the prisoner was at the bar, the dog was let loose in the court-house, and in the midst of some hundreds he found him out (though dressed entirely in new clothes), and would have torn him to pieces had he been allowed; in consequence of which he was condemned, and at the place of execution he confessed the fact. Surely so useful, so disinterestedly faithful an animal, should not be so barbarously treated as I have often seen them, particularly in London."

The following anecdote has been well authenticated, and the fact which it records is still remembered by many individuals yet alive:—

Mr. Alderman Yearsley, of Congleton, in Cheshire, had a favourite large water-spaniel named Fanny, which, in the hands of Providence, was the instrument of saving a very valuable life.

In the year 1774 Mr. Yearsley had gone out one evening with a friend to a tavern, and the dog accompanied him. A short time before he was expected home, and while Mrs. Yearsley happened to be washing her hands in the back kitchen, the spaniel returned and scratched at the door for admittance. Being let Page 322 in, she followed her mistress into the kitchen, where she set up a strange sort of whining, or barking, and turned towards the street-door, as if beckoning her mistress to follow. This she repeated several times, to the great astonishment of the lady. At length a thought struck her that Mr. Yearsley might have met with some accident in the street, and that the spaniel was come to guide her to her husband. Alarmed at this idea, she hastily followed the animal, which led her to Mr. Yearsley, whom she found in perfect health, sitting in the house to which he had gone. She told him the cause of her coming, and got herself laughed at for her pains. But what were the feelings of both, when they were informed by their next neighbours that the kitchen fell in almost the very instant Mrs. Yearsley had shut the street-door, and that the wash-hand basin she had left was crushed into a thousand pieces! The animal was ever afterwards treated with no ordinary attention, and died thirteen years later, at the age of sixteen. Her death, we regret to add, was occasioned by the bite of a mad dog.

In the "Notes of a Naturalist," published in Chambers' "Edinburgh Journal," a work which cannot be too much commended for its agreeable information, is the following anecdote, which I give with the remarks of the author upon it:—

"It appears to me, that in the general manifestations of the animal mind, some one of the senses is employed in preference to the others—that sense, for Page 323 instance, which is most acute and perfect in the animal. In the dog, for example, the sense of smell predominates; and we accordingly find that, through the medium of this sense, his mental faculties are most commonly exercised. A gentleman had a favourite spaniel, which for a long time was in the habit of accompanying him in all his walks, and became his attached companion. This gentleman had occasion to leave home, and was absent for more than a year, during which time he had never seen the dog. On his return along with a friend, while yet at a little distance from the house, they perceived the spaniel lying beside the gate. He thought that this would be a good opportunity of testing the memory of his favourite; and he accordingly arranged with his companion, who was quite unknown to the dog, that they should both walk up to the animal, and express no signs of recognition. As they both approached nearer, the dog started up, and gazed at them attentively; but he discovered no signs of recognition, even at their near approach. At last he came up to the stranger, put his nose close to his clothes, and smelt him, without any signs of emotion. He then did the same to his old master; but no sooner had he smelt him, than recognition instantly took place; he leaped up to his face repeatedly, and showed symptoms of the most extravagant joy. He followed him into the house, and watched his every movement, and could by no means be diverted from his person. Here was an instance of deficient me Page 324 mory through the organs of sight, but an accurate recollection through the organs of smell." In a preceding anecdote, I have recorded an instance of a spaniel recognising the voice of his master after a lapse of six years. In that case, it was evident that the recollection of a particular sound enabled the dog to know his master, without having had recourse to the sense of smelling, which, however, would probably have been equally available had it been exercised.

About the year 1800, Mrs. Osburn, who lived a few miles out of London, went to town to receive a large sum of money granted her by Parliament for discovering a lithontryptic medicine. She received the money, and returned back with it in her own carriage to the country, without anything particular happening to her on the road. It was evening when she arrived at home; and being fatigued with her journey, she retired early to rest. On her stepping into bed, she was somewhat surprised at the importunities of a small King Charles's dog, which was a great pet, and always slept in her bedchamber. He became exceedingly troublesome, and kept pulling the bedclothes with all his strength. She chid him repeatedly, and in an angry tone of voice desired him to lie still, that she might go to sleep. The dog, however, still persisted in his efforts, and kept pulling the bedclothes; and at length leaped on the bed, and endeavoured with the most determined perseverance to pull off the bedclothes. Mrs. Osburn then conceived there must be Page 325 some extraordinary cause for this unusual conduct on the part of her dog, and leaped out of bed; and being a lady of some courage, put on her petticoat, and placed a brace of pistols by her side, which she had always ready loaded in a closet adjoining her bed-room, and proceeded down-stairs. When she had reached the first landing-place, she saw her coachman coming down the private staircase, which led to the servants' rooms, with a lighted candle in his hand, and full dressed. Suspecting his intentions were bad, and with heroic presence of mind, she presented one of her pistols, and threatened to lodge the contents of it in him, unless he returned to bed forthwith. Subdued by her determined courage, he quietly and silently obeyed. She then went into a back-parlour, when she heard a distant whispering of voices; she approached the window, and threw it up, and fired one of her pistols out of it, in the direction from which the noise proceeded. Everything became silent, and not a whisper was to be heard. After looking through the different rooms on the lower floor, and finding all right, she proceeded to bed and secured the door, and nothing further occurred that night. Next morning she arose at an early hour, went into the garden, and in the direction which she had fired the preceding night she discovered drops of blood, which she traced to the other end of the garden. This left no doubt on her mind of what had been intended. Thinking it imprudent to keep so large a sum of money in her Page 326 house, she ordered her carriage to drive to town, where she deposited her cash. She then repaired to the house of Sir John Fielding, and related to him the whole affair, who advised her to part with her coachman immediately, and that he would investigate the matter, and, if possible, discover and convict the offenders. But the parties concerned in this affair were never discovered; for the mere fact of the coachman being found coming down the stair was not sufficient to implicate him, although there were strong grounds of suspicion. Thus, by the instinct and fidelity of this little animal, was robbery, and most likely murder, prevented.

A spaniel belonging to a medical gentleman, with whom I am acquainted, residing at Richmond in Surrey, was in the habit of accompanying him when he went out at night to visit his patients. If he was shut out of the house of a patient, as was frequently the case, he would return home; and whatever the hour of the night might be, he would take the knocker in his mouth, and knock till the door was opened. It should be mentioned that the knocker was below a half-glazed door, so that it was easily within the dog's reach.

"In the capital of a German principality," says Capt. Brown, "the magistrates once thought it expedient to order all dogs that had not the mark of having been wormed, to be seized and confined for a certain time in a large yard without the walls of the town. Page 327 These dogs, which were of all possible varieties, made a hideous noise while thus confined together; but a spaniel, which, as the person that had the care of them observed, sat apart from the rest in a corner of the yard, seemed to consider the circumstances with greater deliberation. He attended to the manner in which the gate of the yard was opened and shut; and, taking a favourable opportunity, leapt with his forepaws upon the latch, opened the gate, looked round upon the clamorous multitude, and magnanimously led them the way out of the prison. He conducted them in triumph through the gate of the town; upon which every dog ran home exulting to his master."

The following anecdote, which was sent to me by the gentleman who witnessed the occurrence, proves the sense and observation of a spaniel. He possessed one which was a great favourite, and a constant companion in all his rambles. One day, in passing through a field of young turnips, he pulled up one of them, and after washing it carefully in a rivulet, he cut off the top, and ate the other part. During this time the dog eyed him attentively, and then proceeded to one of the growing turnips, drew it from the earth, went up briskly to the rivulet, and after dashing it about some time till he caused the water to froth considerably, he laid it down, and holding the turnip inverted, and by the top, he deliberately gnawed the whole of it off, and left the top, thus closely imitating the actions of his master.

Page 328 A gentleman, who generally resided at Boston in Lincolnshire, had also a house at Chepstow in Monmouthshire, to which he occasionally went in the summer. While at the latter place, a small spaniel dog which a friend at Chepstow had given him was taken on his return in a carriage to Boston. On the Sunday evening after the arrival at that place, the spaniel was attacked by a large dog, when out walking with his master on the river bank, and ran away. Nothing was heard of him until the receipt of a letter from Chepstow, announcing his arrival at that place in a famished and travel-worn condition. The distance is one hundred and eighty-four miles.

The following anecdote is related by Mr. Blaine:—

"I was once called from dinner in a hurry to attend to something that had occurred; unintentionally I left a favourite cat in the room, together with a no less favourite spaniel. When I returned I found the latter, which was not a small figure, extending her whole length along the table by the side of a leg of mutton which I had left. On my entrance she showed no signs of fear, nor did she immediately alter her position. I was sure, therefore, that none but a good motive had placed her in this extraordinary situation, nor had I long to conjecture. Puss was skulking in a corner, and though the mutton was untouched, yet her conscious fears clearly evinced that she had been driven from the table in the act of attempting a robbery on the meat, to which she was too prone, and that her Page 329 situation had been occupied by this faithful spaniel to prevent a repetition of the attempt. Here was fidelity united with great intellect, and wholly free from the aid of instinct. This property of guarding victuals from the cat, or from other dogs, was a daily practice of this animal; and, while cooking was going forward, the floor might have been strewed with eatables, which would have been all safe from her own touch, and as carefully guarded from that of others. A similar property is common to many dogs, but to spaniels particularly."

It is impossible in a work on dogs to omit the insertion of some pretty lines on a spaniel by Mrs. Barrett Browning, and which do so much credit to her kindly feelings and poetic talents:—

"Yet, my pretty sportive friend,
Little is't to such an end
That I praise thy rareness!
Other dogs may be thy peers,
Haply, in those drooping ears,
And this glossy fairness.

But of thee it shall be said,
'This dog watched beside a bed
Day and night unweary,—
Watched within a curtained room
Where no sunbeam broke the gloom
Round the sick and dreary.

Roses, gathered for a vase,
In that chamber died apace,
Beam and breeze resigning—
This dog only waited on,
Knowing that when light is gone

Love remains for shining.

Page 330

Other dogs, in thymy dew,
Tracked the hares and followed through
Sunny moor or meadow—
This dog only crept and crept
Next a languid cheek that slept,
Sharing in the shadow.

Other dogs of loyal cheer
Bounded at the whistle clear,
Up the woodside hieing—
This dog only watched in reach
Of a faintly uttered speech,
Or a louder sighing.

And if one or two quick tears
Dropped upon his glossy ears,
Or a sigh came double,—
Up he sprang in eager haste,
Fawning, fondling, breathing fast,
In a tender trouble.'"

Page 331

FRENCH POODLE.

THE POODLE.

"With all the graces of his fatherland;
With well-cut coat, and ever ready hand—
See—the French poodle sports his life away;
Obedient, wise, affectionate, and gay."—*Chronicles of Animals.*

These dogs, like all others, possess many amiable qualities, and are remarkable for the facility with which they learn several amusing tricks, and for their extraordinary sagacity. This latter quality has frequently made them a great source of profit to their masters, so that it may be said of them, "c'est encore une des plus profitables manières d'être chien qui existent." A proof of this is related by M. Blaze in his history of the dog, and was recorded by myself many years before his work appeared.

Page 332 A shoe-black on the Pont Neuf at Paris had a poodle dog, whose sagacity brought no small profit to his master. If the dog saw a person with well-polished boots go across the bridge, he contrived to dirty them, by having first rolled himself in the mud of the Seine. His master was then employed to clean them. An English gentleman, who had suffered more than once from the annoyance of having his boots dirtied by a dog, was at last induced to watch his proceedings, and thus detected the tricks he was playing for his master's benefit. He was so much pleased with the animal's sagacity, that he purchased him at a high price and conveyed him to London. On arriving there, he was confined to the house till he appeared perfectly satisfied with his new master and his new situation. He at last, however, contrived to escape, and made his way back to Paris, where he rejoined his old master, and resumed his former occupation. I was at Paris some years ago, where this anecdote was related to me, and it is now published in the records of the French Institute.

Perhaps the most remarkable instance known of what are called "Learned Dogs," is that of two poodles, which were trained at Milan, and exhibited at Paris in the spring of 1830. The account of them is given by a lady, whose veracity is not doubtful, and who herself saw their performance. "The elder, named Fido," says she, "is white, with some black patches on his head and back; and the younger, who Page 333 is called Bianco, is also white, but with red spots. Fido is a grave and serious personage, walks with dignity round the circle assembled to see him, and appears much absorbed in reflection. Bianco is young and giddy, but full of talent when he chooses to apply it. Owing to his more sedate disposition, however, Fido is called upon to act the principal part of the exhibition. A word is dictated to him from the Greek, Latin, Italian, German, French, or English language, and selected from a vocabulary where fifty words in each tongue are inscribed, and which all together make three hundred different combinations. An alphabet is placed before Fido, and from it he takes the letters which compose the given word, and lays them in proper order at the feet of his master. On one occasion he was told to spell the word Heaven, and he quickly placed the letters till he came to the second e; he stood for an instant as if puzzled, but in a moment after he took the e out of the first syllable, and put it into the second. His attainments in orthography, however, are not so surprising as those in arithmetic. He practises the four rules with extraordinary facility, arranges the double ciphers as he did the double vowels in the word Heaven, and rarely makes an error. When such does occur, his more thoughtless companion is called in to rectify it, which he invariably does with the greatest quickness; but as he had rather play than work, and pulls Fido by the ears to make him as idle as himself, he is quickly dismissed. One Page 334 day, the steady Fido spelt the word Jupiter with a *b* instead of a *p*; Bianco was summoned to his aid, who, after contemplating the word, pushed out the *b* with his nose, and seizing a *p* between his teeth, put it into the vacancy. Fido is remarkable for the modest firmness with which he insists upon his correctness when he feels convinced of it himself; for a lady having struck a repeating watch in his ear, he selected an 8 for the hour, and a 6 for the three-quarters. The company present, and his master, called out to him he was wrong. He reviewed his numbers and stood still. His master insisted, and he again examined his ciphers; after which he went quietly, but not in the least abashed, into the middle of the carpet, and looked at his audience. The watch was then sounded again, and it was found to have struck two at every quarter; and Fido received the plaudits which followed with as gentle a demeanour as he had borne the accusation of error.

"One occupation seems to bring the giddy Bianco to the gravity of the elder savant; and when the spectators are tired of arithmetic and orthography, the two dogs either sit down to *écarté*, or become the antagonists of one of the company. They ask for, or refuse cards, as their hands require, with a most important look; they cut at the proper times, and never mistake one suit for another. They have recourse to their ciphers to mark their points; and on one occasion Bianco having won, he selected his number, and on Page 335 being asked what were the gains of his adversary, he immediately took an O between his teeth, and showed it to the querist; and both seemed to know all the terms of the game as thoroughly as the most experienced card-players. All this passes without the slightest visible or audible sign between the poodles and their master; the spectators are placed within three steps of the carpet on which the performance goes forward; people have gone for the sole purpose of watching the master; everybody visits them, and yet no one has hitherto found out the mode of communication established between them and their owner. Whatever this communication may be, it does not deduct from the wonderful intelligence of these animals; for there must be a multiplicity of signs, not only to be understood with eyes and ears, but to be separated from each other in their minds, or to be combined one with another, for the various trials in which they are exercised.

"I have seen learned pigs and ponies, and can, after these spectacles, readily imagine how the extraordinary sagacity of a dog may be brought to a knowledge of the orthography of three hundred words; but I must confess myself puzzled by the acquirements of these poodles in arithmetic, which must depend upon the will of the spectator who proposes the numbers; but that which is most surprising of all is the skill with which they play *écarté*. The gravity and attention with which they carry on their game is almost ludi Page 336 crous; and the satisfaction of Bianco when he marks his points is perfectly evident."

Nor is this a solitary instance of the extraordinary sagacity of the poodle. A lady of my acquaintance had one for many years, who was her constant companion both in the house and in her walks. When, however, either from business or indisposition, her mistress did not take her usual walk on Wimbledon Common, the dog, by jumping on a table, took down the maid-servant's bonnet, and held it in her month till she accompanied the animal to the Common.

A friend of mine had a poodle dog, who was not very obedient to his call when he was taken out to run in the fields. A small whip was therefore purchased, and the dog one day was chastised with it. The whip was placed on a table in the hall of the house, and the next morning it could not be found. It was soon afterwards discovered in the coal-cellar. The dog was a second time punished with it, and again the whip was missed. It was afterwards discovered that the dog had attempted to hide the instrument by which pain had been inflicted on him. There certainly appears a strong approach to reason in this proceeding of the dog. *Cause* and *effect* seem to have been associated in his mind, if his mode of proceeding may be called an effort of it.

In Messrs. Chambers' brochure of amusing anecdotes of dogs we find the following: —

An aged gentleman has mentioned to us that, Page 337 about fifty years ago, a Frenchman brought to London from eighty to a hundred dogs, chiefly poodles, the remainder spaniels, but all nearly of the same size, and of the smaller kind. On the education of these animals their proprietor had bestowed an immense deal of pains. From puppyhood upwards they had been taught to walk on their hind-legs, and maintain their footing with surprising ease in that unnatural position. They had likewise been drilled into the best possible behaviour towards each other; no snarling, barking, or indecorous conduct took place when they were assembled in company. But what was most surprising of all, they were able to perform in various theatrical pieces of the character of pantomimes, representing various transactions in heroic and familiar life, with wonderful fidelity. The object of their proprietor was, of course, to make money by their performances, which the public were accordingly invited to witness in one of the minor theatres.

Amongst their histrionic performances was the representation of a siege. On the rising of the curtain there appeared three ranges of ramparts, one above the other, having salient angles and a moat, like a regularly-constructed fortification. In the centre of the fortress arose a tower, on which a flag was flying; while in the distance behind appeared the buildings and steeples of a town. The ramparts were guarded by soldiers in uniform, each armed with a musket or sword, of an appropriate size. All these were dogs, Page 338 and their duty was to defend the walls from an attacking party, consisting also of dogs, whose movements now commenced the operations of the siege. In the foreground of the stage were some rude buildings and irregular surfaces, from among which there issued a reconnoitring party; the chief, habited as an officer of rank, with great circumspection surveyed the fortification; and his sedate movements, and his consultations with the troops that accompanied him, implied that an attack was determined upon. But these consultations did not pass unobserved by the defenders of the garrison. The party was noticed by a sentinel and fired upon; and this seemed to be the signal to call every man to his post at the embrasures.

Shortly after, the troops advanced to the escalade; but to cross the moat, and get at the bottom of the walls, it was necessary to bring up some species of pontoon, and, accordingly, several soldiers were seen engaged in pushing before them wicker-work scaffoldings, which moved on castors, towards the fortifications. The drums beat to arms, and the bustle of warfare opened in earnest. Smoke was poured out in volleys from shot-holes; the besieging forces pushed forward in masses, regardless of the fire; the moat was filled with the crowd; and, amid much confusion and scrambling, scaling-ladders were raised against the walls. Then was the grand tug of war. The leaders of the forlorn hope who first ascended were opposed with great gallantry by the defenders; and this was, Page 339 perhaps, the most interesting part of the exhibition. The chief of the assailants did wonders; he was seen now here, now there, animating his men, and was twice hurled, with ladder and followers, from the second gradation of ramparts: but he was invulnerable, and seemed to receive an accession of courage on every fresh repulse. The rattle of the miniature cannon, the roll of the drums, the sound of trumpets, and the heroism of the actors on both sides, imparted an idea of reality to the scene.

After numerous hairbreadth escapes, the chief surmounted the third line of fortifications, followed by his troops; the enemy's standard was hurled down, and the British flag hoisted in its place; the ramparts were manned by the conquerors; and the smoke cleared away, to the tune of "God save the King."

It is impossible to convey a just idea of this performance, which altogether reflected great credit on its contriver, as also on the abilities of each individual dog. We must conclude that the firing from the embrasures, and some other parts of the *méchanique*, were effected by human agency; but the actions of the dogs were clearly their own, and showed what could be effected with animals by dint of patient culture.

Another specimen of these canine theatricals was quite a contrast to the bustle of the siege. The scene was an assembly-room, on the sides and the further end of which seats were placed; while a music-gallery, and a profusion of chandeliers, gave a richness and Page 340 truth to the general effect. Livery-servants were in attendance on a few of the company, who entered and took their seats. Frequent knockings now occurred at the door, followed by the entrance of parties attired in the fashion of the period. These were, of course, the same individuals who had recently been in the deadly breach; but now all was tranquillity, elegance, and ease. Parties were formally introduced to each other with an appearance of the greatest decorum. The dogs intended to represent ladies were dressed in silks, gauzes, laces, and gay ribbons. Some wore artificial flowers, with flowing ringlets; others wore the powdered and pomatumed head-dress, with caps and lappets, in ludicrous contrast to the features of the animals. The animals which represented gentlemen were judiciously equipped; some as youthful and others as aged beaux, regulated by their degrees of proficiency, since those most youthfully dressed were most attentive to the ladies. The frequent bow and return of curtsey produced great mirth in the audience. On a sudden the master of the ceremonies appeared; he wore a superb court-dress, and his manners were in agreement with his costume. To some of the gentlemen he gave merely a look of recognition; to the ladies he was generally attentive; to some he projected his paw familiarly, to others he bowed with respect; and introduced one to another with an air of elegance that surprised and delighted the spectators.

As the performance advanced the interest increased. Page 341 The music was soon interrupted by a loud knocking, which announced the arrival of some important visitor. Several livery servants entered, and then a sedan-chair was borne in by appropriately dressed dogs; they removed the poles, raised the head, and opened the door of the sedan; forth came a lady, splendidly attired in spangled satin and jewels, and her head decorated with a plume of ostrich feathers! She made a great impression, and appeared as if conscious of her superior attraction; meanwhile the chair was removed, the master of the ceremonies, in his court-dress, was in readiness to receive the *élégante,* and the bow and curtsey were admirably interchanged. The band now struck up an air of the kind to which ball-room companies are accustomed to promenade, and the company immediately quitted their seats and began to walk ceremoniously in pairs round the room. Three of the ladies placed their arms under those of their attendant gentlemen. On seats being resumed, the master of the ceremonies and the lady who came in the sedan-chair arose; he led her to the centre of the room; Foote's minuet struck up; the pair commenced the movements with an attention to time; they performed the crossings and turnings, the advancings, retreatings, and obeisances, during which there was a perfect silence, and they concluded amid thunders of applause. What ultimately became of the ingenious manager with his company, our informant never heard.

The following anecdotes prove the strong affection Page 342 and perseverance of the poodle. The late Duke of Argyll had a favourite dog of this description, who was his constant companion. This dog, on the occasion of one of the Duke's journeys to Inverary Castle, was, by some accident or mistake, left behind in London. On missing his master, the faithful animal set off in search of him, and made his way into Scotland, and was found early one morning at the gate of the castle. The anecdote is related by the family, and a picture shown of the dog.

A poor German artist, who was studying at Rome, had a poodle dog, who used to accompany him, when his funds would allow it, to an ordinary frequented by other students. Here the dog got scraps enough to support him. His master, not being able to keep up the expense, discontinued his visits to the ordinary. The dog fared badly in consequence, and at last his master returned to his friends in Germany, leaving the dog behind him. The poor animal slept at the top of the stairs leading to his master's room, but watched in the day time at the door of the ordinary, and when he saw his former acquaintances crowding in, he followed at their heels, and thus gaining admittance was fed till his owner came back to resume his studies.

A gentleman possessed a poodle dog and a terrier, between whom a great affection existed. When the terrier was shut up, as was sometimes the case, the poodle always hid such bones or meat as he could procure, and afterwards brought the terrier to the spot Page 343 where they were concealed. He was constantly watched, and observed to do this act of kindness.

The sagacity of the poodle is strongly shown by the following fact. Mr. B— —t, who was constantly in the habit of making tours on the Continent, was always accompanied by a poodle dog. In one of his journeys he was seated at a table-d'hôte next to a person whose conversation he found so agreeable, that a sort of intimacy sprung up between them. The dog, however, for the first time he had ever done so to any one, showed a dislike to the stranger, and so much so, that Mr. B— —t could not help remarking it. In the course of his tour he again fell in with the stranger, when the intimacy was renewed, and Mr. B— —t offered him a seat in his carriage as they were both going the same way. No sooner, however, had the stranger entered the carriage, than the dog showed an increased dislike of him, which continued during the course of the journey. At night they slept at a small inn, in a wild and somewhat unfrequented country, and on separating in the evening to go to their respective beds, the poodle evinced the greatest anger, and was with difficulty restrained from attacking the stranger. In the middle of the night Mr. B— —t was awoke by a noise in his room, and there was light enough for him to perceive that his dog had seized his travelling companion, who, upon being threatened, confessed that he had entered the room for the purpose of Page 344 endeavouring to purloin Mr. B— —t's money, of which he was aware that he possessed a considerable quantity. This is not a solitary instance of an instinctive faculty which enables dogs to discriminate, by showing a strong dislike, the characters of particular individuals.

A friend has sent me the following account of a poodle he once had:—

"Many years ago I had a poodle who was an excellent retriever. He was a middle-sized, active dog, a first-rate waterman, with a nose so particularly sensitive that no object, however minute, could escape its 'delicate investigation.' Philip was the hardiest animal in the world—no sea would prevent him from carrying a dead bird through the boiling breakers, and I have seen him follow and secure a wounded mallard, although in the attempt his legs were painfully scarified in breaking through a field of ice scarcely the thickness of a crown-piece. Philip, though of French extraction, had decidedly Irish partialities. He delighted in a glass of grog; and no matter with what labour and constancy he had returned from retrieving, he still enjoyed a glass of punch. When he had drunk it, he was in high glee, running round and round to try and catch his own tail, and even then allowing the cat to approach him, which he was by no means disposed to do at other times."

When my daughter was in Germany, she sent me the following interesting anecdote of a poodle, the accuracy of which she had an opportunity of ascertaining.

Page 345 An inhabitant of Dresden had a poodle that he was fond of, and had always treated kindly. For some reason or another he gave her to a friend of his, a countryman in Possenderf, who lived three leagues from Dresden. This person, who well knew the great attachment of the dog to her former master, took care to keep her tied up, and would not let her leave the house till he thought she had forgotten him. During this time the poodle had young ones, three in number, which she nourished with great affection, and appeared to bestow upon them her whole attention, and to have entirely given up her former uneasiness at her new abode. From this circumstance her owner thought she had forgotten her old master, and therefore no longer kept her a close prisoner. Very soon, however, the poodle was missing, and also the three young ones, and nothing was heard of her for several days. One morning his friend came to him from Dresden, and informed him that the preceding evening the poodle had come to his house with one of the puppies in her mouth, and that another had been found dead on the road to Possenderf. It appeared that the dog had started in the night, carrying the puppies (who were not able to walk) one after the other, a certain distance on the road to Dresden, with the evident intention of conveying them all to her much-loved home and master. The third puppy was never found, and is supposed to have been carried off by some wild animal or bird, while Page 346 the poor mother was in advance with the others. The dead one had apparently perished from cold.

The late Dr. Chisholm of Canterbury had a remarkable poodle, which a correspondent informs me he has often seen. On one occasion he was told, for the first time, by way of trial, to fetch his master's slippers. He went up-stairs, and brought down one only. He was then told, "You have brought one only, go and fetch the other;" and the other was brought. The next evening the dog was again told to bring the slippers. He went up-stairs, put one slipper within the other, and brought both down. This dog appeared to understand much of our language. When dining with Dr. Chisholm and others, his intelligence was put to the proof by my correspondent. Some one would hide an article, open the door, and bring in the dog, saying, "Find so-and-so." The poodle used to look up steadily in the face of the speaker, until he was told whether the article was hid high or low; he would then search either on the ground, or on the chairs and furniture, and bring the article, never taking any notice of any other thing that was lying about. He would, upon being ordered, go up-stairs and bring down a snuff-box, stick, pocket-handkerchief, or anything, understanding as readily what was said to him as if spoken to a servant.

Another poodle would go through the agonies of dying in a very systematic manner. When he was ordered to die, he would tumble over on one side, and Page 347 then stretch himself out, and move his hind legs in such a way as expressed that he was in great pain, first slowly and afterwards very quickly. After a few convulsive throbs, indicated by putting his head and whole body in motion, he would stretch out all his limbs and cease to move, lying on his back with his legs turned upwards, as if he had expired. In this situation he remained motionless until he had his master's commands to get up.

The following anecdote was communicated to the Rev. Mr. Jenyns by Mrs. Grosvenor, of Richmond, Surrey:—

A poodle dog belonging to a gentleman in Cheshire was in the habit of not only going to church, but of remaining quietly in the pew during service, whether his master was there or not. One Sunday the dam at the head of a lake in that neighbourhood gave way, so that the whole road was inundated. The congregation, in consequence, consisted of a very few, who came from some cottages close by, but nobody attended from the great house. The clergyman informed the lady, that whilst reading the Psalms he saw his friend, the poodle, come slowly up the aisle dripping with wet, having swam above a quarter of a mile to get to church. He went into the usual pew, and remained quietly there to the end of the service.

The Marquess of Worcester (the late Duke of Beaufort), who served in the Peninsular war, had a poodle which was taken from the grave of his master, Page 348 a French officer, who fell at the battle of Salamanca, and was buried on the spot. The dog had remained on the grave until he was nearly starved, and even then was removed with difficulty; so faithful are these animals in protecting the remains of those they loved.

A poodle dog followed his master, a French officer, to the wars; the latter was soon afterwards killed at the battle of Castella, in Valencia, when his comrades endeavoured to carry the dog with them in their retreat; but the faithful animal refused to leave the corpse, and they left him. A military marauder, in going over the field of battle, discovering the cross of the legion of honour on the dead officer's breast, attempted to capture it, but the poodle instantly seized him by the throat, and would have ended his career had not a comrade run the honest canine guardian through the body.

Mr. Blaine, in his "Account of Dogs," says that, "strange as it may appear, it is no less true, that a poodle dog actually scaled the high buildings of my residence in Wells Street, Oxford Street, proceeded along several roofs of houses, and made his way down by progressive but very considerable leaps into distant premises; from whence, by watching and stratagem, he gained the street, and returned home in order to join his mistress, for whose sake he had encountered these great risks."

I am always glad to have an opportunity of acknowledging the kindness of my correspondents, and now Page 349 do so to the clergyman who very kindly sent me the following anecdote, which I give in his own words:—

"I have a distinct remembrance of Froll or Frolic, a dog belonging to an aged relation, once the property of her deceased only son, which animal, in his earlier days, doubtless gave evidence that his name was not given him unadvisedly, but during the yearly visits of myself to that kind and indulgent person, I can remember nothing but a rather small though fat unwieldy poodle, whose curly, glossy coat (preserved after his death), long yellow ears, and black nose, the rest of his body being perfectly white, betokened that he had been a beauty in his time. Froll was still a prodigious favourite with his mistress, although I confess my feelings towards him were rather those of fear than any other, for to touch him was quite sufficient to evoke a growl, or perchance a snap, from this pet of a dozen years or more. A cross, snappish fellow he was at best, and well he knew the length of Trusty the house-dog's chain, which less favoured quadruped was never let loose by day, from a well-grounded fear that he might, if allowed, resent, by summary punishment, the constant insults he was doomed to submit to from this most petted and presumptuous myrmidon of the drawing-room. With all this, although time and over-feeding had soured his temper, Froll still retained much of, if not all, his former intelligence (a trait so peculiar to his species), declared by many long-past but still vaunted proofs of his being a wonder in his way. Page 350 One of his peculiarities was a fondness for apples—not indeed all apples, but those which grew on a particular tree, called 'Froll's tree,' and no others; this tree was, by the way, the best in the garden, and the small, sweet, delicate fruit therefrom (my reminiscence is distinct on this point) were carefully preserved for this canine favourite. Nothing would entice him to eat any other sort of apple. And in the season he would constantly urge his mistress into the garden by repeated barking, and other unmistakable symptoms. His daily meals, too, of which I think there were three regular ones, were events in themselves, the careful attention to which tended perhaps to relieve the monotony of a country life: they are indeed not speedily to be forgotten by those who witnessed them. He would take food from no one but his mistress or her maid, which latter person was his chief purveyor, who had been an inmate of the house contemporary with himself, or I believe long before; but this feeding was generally a task of great trouble, such coaxing and humouring on the one hand, such growling and snarling on the other, has been perhaps seldom heard. At length, after much beseeching on the part of the maid, and a few words of entreaty from the mistress, he would condescend to eat; but never, I believe, without some symptoms of discontent, how savoury soever the morsel, submitting to that as a favour which is generally snatched at and devoured with so much gusto and avidity by most others of his tribe. I should not have entered into these peculi Page 351 arities, which are scarcely evidence of any intelligence beyond that of other dogs, were it not that the circumstances attending his death were really extraordinary, the more so when the character of the dog is considered; and as we have so often heard of a presentiment of that great change being strongly imprinted on human minds, so there were not wanting some of the then inmates of the house, who attributed his unwonted behaviour on the eve of his death to the same cause. The dog slept constantly in his mistress's bed-room, but, contrary to custom on the night in question, he pertinaciously refused to remain there. My brother and myself, who were then little boys, were, to our great surprise, aroused in the course of the night by an unwonted scratching at the door of our apartment, which we immediately opened, and, to our equal delight and wonder, were saluted by Froll's jumping up and licking our hands and faces—certainly he never appeared in better health and spirits in his life. Whether he did this to atone for his former uncourteous behaviour towards us, or was urged by some unaccountable feeling of amiability as well as restlessness, I cannot say, but certain it is his gentler faculties were that night for once aroused, for this unaccustomed compliment I can safely affirm we never personally received at any former period of our acquaintance. After a time he left us, charmed at experiencing these new and flattering demonstrations; which joy was, alas! doomed to be sadly and speedily extinguished. When the Page 352 morning came, the distressed countenance of the servant who called us, portended some evil tidings, which was quickly followed by the unexpected intelligence of the demise of poor Froll. We hastily accompanied the servant into the coachman's sleeping apartment, and there, under the bed, lay the poor dog. It had pleased him to go there to die, having previously aroused every individual in the house during the night by scratching at their several chambers one after another, and saluting them in the same amiable manner he had my brother and myself."

This anecdote could be well authenticated by most of the persons then in the house, who are still alive.

Page 353

THE ESQUIMAUX DOG.

Dr. Richardson, in his "American Fauna," mentions as a curious fact, that those Indian nations who still preserve their ancient mode of life, have dogs which bear a strong resemblance to wolves. Thus it is with the Esquimaux dogs. They are extremely like the grey wolves of the Arctic Circle in form and colour, and nearly equal to them in size. They also bear some resemblance to the Pomeranian breed, although the latter are much smaller.

It is curious that almost every nation on earth has some particular traditions regarding the dog. The Esquimaux, a nation inhabiting the polar regions, have a singular fable amongst them respecting the origin of the Dog-Rib Indians, a tribe which inhabits the northern confines of the American continent. It Page 354 is thus detailed in Captain Franklin's "Second Journey to the Polar Sea:" —

"For a long time Chapawee's descendants were united as one family, but at length some young men being accidentally killed in a game, a quarrel ensued, and a general dispersion of mankind took place. One Indian fixed his residence on the borders of the lake, taking with him a dog big with young. The pups in due time were littered, and the Indian, when he went out to fish, carefully tied them up to prevent their straying. Several times, as he approached his tent, he heard a noise of children talking and playing; but on entering it, he only perceived the pups tied up as usual. His curiosity being excited by the voices he had heard, he determined to watch; and one day pretending to go out and fish, according to custom, he concealed himself in a convenient place. In a short time he again heard voices, and rushing suddenly into the tent, beheld some beautiful children sporting and laughing, with the dog-skins lying by their side. He threw the skins into the fire, and the children, retaining their proper forms, grew up, and were the ancestors of the Dog-Rib nation."

Captain Lyon, who had so many opportunities of studying the habits of the Esquimaux dog, has given so interesting an account of it that I cannot do better than quote his own words: —

"Having myself possessed, during our hard winter, a team of eleven fine dogs, I was enabled to become Page 355 better acquainted with their good qualities than could possibly have been the case by the casual visits of the Esquimaux to the ships. The form of the Esquimaux dog is very similar to that of our shepherds' dog in England, but it is more muscular and broad-chested, owing to the constant and severe work to which he is brought up. His ears are pointed, and the aspect of the head is somewhat savage. In size a fine dog is about the height of the Newfoundland breed, but broad like a mastiff in every part except the nose. The hair of the coat is in summer, as well as in winter, very long, but during the cold season a soft, downy under-covering is found, which does not appear in warm weather. Young dogs are put into harness as soon as they can walk, and being tied up, soon acquire a habit of pulling, in their attempts to recover their liberty, or to roam in quest of their mother. When about two months old, they are put into the sledge with the grown dogs, and sometimes eight or ten little ones are under the charge of some steady old animal, where, with frequent and sometimes severe beatings, they soon receive a competent education. Every dog is distinguished by a particular name, and the angry repetition of it has an effect as instantaneous as an application of the whip, which instrument is of an immense length, having a lash from eighteen to twenty-four feet, while the handle is one foot only; with this, by throwing it on one side or the other of the leader, and repeating certain words, the animals are guided or Page 356 stopped. When the sledge is stopped they are all taught to lie down, by throwing the whip gently over their backs, and they will remain in this position even for hours, until their master returns to them. A walrus is frequently drawn along by three or four of these dogs, and seals are sometimes carried home in the same manner, though I have in some instances seen a dog bring home the greater part of a seal in panniers placed across his back. The latter mode of conveyance is often used in summer, and the dogs also carry skins or furniture overland to the sledges when their masters are going on any expedition. It might be supposed that in so cold a climate these animals had peculiar periods of gestation, like the wild creatures, but, on the contrary, they bear young at every season of the year, and seldom exceed five at a litter. Cold has very little effect on them; for although the dogs at the huts slept within the snow passages, mine at the ships had no shelter, but lay alongside, with the thermometer at 42° and 44°, and with as little concern as if the weather had been mild. I found, by several experiments, that three of my dogs could draw me on a sledge, weighing one hundred pounds, at the rate of one mile in six minutes; and as a proof of the strength of a well-grown dog, my leader drew one hundred and ninety-six pounds singly, and to the same distance, in eight minutes. At another time seven of my dogs ran a mile in four minutes, drawing a heavy sledge full of men. Afterwards, in carrying stores to the Fury, one mile distant, Page 357 nine dogs drew one thousand six hundred and eleven pounds in the space of nine minutes. My sledge was on runners, neither shod nor iced; but had the runners been iced, at least forty pounds weight would have been added for each dog."

Captain Lyon, in another passage, observes:—"Our eleven dogs were large, and even majestic-looking animals; and an old one of peculiar sagacity was placed at their head by having a longer trace, so as to lead them through the safest and driest places, these animals having such a dread of water as to receive a severe beating before they would swim a foot. The leader was instant in obeying the voice of the driver, who never beat, but repeatedly called to him by name. When the dogs slackened their pace, the sight of a seal or bird was sufficient to put them instantly to their full speed; and even though none of these might be seen on the ice, the cry of "a seal!"—"a bear!"—or "a bird!" &c., was enough to give play to the legs and voices of the whole pack. It was a beautiful sight to observe the two sledges racing at full speed to the same object, the dogs and men in full cry, and the vehicles splashing through the holes of water with the velocity and spirit of rival stage-coaches. There is something of the spirit of professed whips in these wild races, for the young men delight in passing each other's sledge, and jockeying the hinder one by crossing the path. In passing on different routes the right hand is yielded, and should an inexperienced driver Page 358 endeavour to take the left, he would have some difficulty in persuading his team to do so. The only unpleasant circumstance attending these races is, that a poor dog is sometimes entangled and thrown down, when the sledge, with perhaps a heavy load, is unavoidably drawn over his body. The driver sits on the fore part of the vehicle, from whence he jumps when requisite to pull it clear of any impediments which may lie in the way, and he also guides it by pressing either foot on the ice. The voice and long whip answer all the purposes of reins, and the dogs can be made to turn a corner as dexterously as horses, though not in such an orderly manner, since they are constantly fighting; and I do not recollect to have seen one receive a flogging without instantly wreaking his passion on the ears of his neighbours. The cries of the men are not more melodious than those of the animals; and their wild looks and gestures when animated, give them an appearance of devils driving wolves before them. Our dogs had eaten nothing for forty-eight hours, and could not have gone over less than seventy miles of ground; yet they returned, to all appearance, as fresh and active as when they first set out."

Such is the Esquimaux dog, an animal of the greatest value in the cold regions of the Arctic circle. In addition to Captain Lyon's very interesting account of them, it may be mentioned that they are of great use to their masters in discovering by the scent the winter retreats which the bears make under the snow. Their en Page 359 durance, too, never tires, and their fidelity is never shaken by blows and starving: they are obstinate in their nature, but the women, who treat them with more kindness than the men, and who nurse them in their helpless state, or when they are sick, have an unbounded command over their affections.

I am indebted to Colonel Hamilton Smith for the following account of an Esquimaux dog brought to this country, and which he received from Mr. Cleghorn, the owner of the animal: —

"The Esquimaux dog is possessed of very great sagacity — in some respects, more than any dog I have ever seen. I may mention an instance. In coming along a country road a hare started, and in place of running after the hare in the usual way, the dog pushed himself through the hedge, crossed the field, and, when past the hare, through the hedge again, as if to meet her direct. It is needless to remark, that the hare doubled through the hedge; but had it been in an open country, there would have been a fine chase. One particular characteristic of the dog is, that he forms a strong attachment to his master, and however kind others may be, they never can gain his affection, even from coaxing with food or otherwise; and, whenever set at liberty, he rushes to the spot where the individual of his attachment is. I may give one or two instances among many. One morning he was let loose by some of the men on the ground, when he instantly bounded from them to my house, and the kitchen-door Page 360 being open, found his way through it; when, to the great amazement of all, he leaped into the bed where I was sleeping, and fawned in the most affectionate manner upon me. Another instance was, when the dog was with me going up the steep bank of the Prince's Street garden, I slipped my foot and came down, when he immediately seized me by the coat, as if to render assistance in raising me. Notwithstanding this particular affection to some, he was in the habit of biting others, without giving the least warning or indication of anger. He was remarkably cunning, for he was in the practice of strewing his meat around him, to induce fowls or rats to come within his reach while he lay watching, as if asleep, when he instantly pounced upon them, and always with success. He was swift, and had a noble appearance when running."

Page 361

THE OTTER TERRIER.

"How greedily
They snuff the fishy steam, that to each blade
Rank scenting clings! See! how the morning dews
They sweep, that from their feet besprinkling drop
Dispersed, and leave a track oblique behind.
Now on firm land they range, then in the flood
They plunge tumultuous; or through reedy pools
Rustling they work their way; no holt escapes
Their curious search. With quick sensation now
The fuming vapour stings; flutter their hearts,
And joy redoubled bursts from every mouth
In louder symphonies. Yon hollow trunk,
That with its hoary head incurv'd salutes
The passing wave, must be the tyrant's fort
And dread abode. How these impatient climb,

While others at the root incessant bay!—
They put him down."—Somerville.

The above is an animated and beautiful description of an otter hunt, an old English sport fast falling into Page 362 disuse, and the breed of the real otter-hound is either extinct or very nearly so. In stating this, I am aware that there are still many dogs which are called otter-hounds; but it may be doubted whether they possess that peculiar formation which belongs exclusively to the true breed. Few things in nature are more curious and interesting than this formation, and it shows forcibly how beautifully everything has been arranged for the instincts and several habits of animals. The true otter-hound is completely web-footed, even to the roots of its claws; thus enabling it to swim with much greater facility and swiftness than other dogs. But it has another extraordinary formation; the ear possesses a sort of flap, which covering the aperture excludes the entrance of the water, and thus the dog is enabled to dive after the otter without that inconvenience which it would otherwise experience. The Earl of Cadogan has, what his Lordship considers, the last of the breed of the true otter-hound. It was a present from Sir Walter Scott. Lord Cadogan offered one hundred pounds for another dog of the same breed, but of a different sex; but I believe without being able to procure one with those true marks which are confined to the authentic breed. A gipsy was, indeed, said to have possessed one, but he refused to part with it.

Those who saw the exhibition of pictures in the Royal Academy in 1844 will recollect a large, interesting, and beautiful picture by Sir Edwin Landseer of a Page 363 pack of otter-hounds. The picture describes the hunt at the time of the termination of the chase and the capture of the otter. The animal is impaled on the huntsman's spear, while the rough, shaggy, and picturesque-looking pack are represented with eyes intently fixed on the amphibious beast, and howling in uncouth chorus round their agonized and dying prey.

An otter-hunt is a cheerful and inspiriting sport, and it is still carried on in some of the lakes of Cumberland. Indeed, as lately as the year 1844, a pack of otter-hounds was advertised in the newspapers to be sold by private contract. The alleged cause of the owner's parting with them was in consequence of their having cleared the rivers of three counties (Staffordshire being one) of all the otters, and the number captured and killed in the last few years was mentioned. "Good otter-hounds," as an old writer observes, "will come chanting, and trail along by the river-side, and will beat every tree-root, every osier-bed, and tuft of bulrushes; nay, sometimes they will take the water and beat it like a spaniel, and by these means the otter can hardly escape you." The otter swims and dives with great celerity, and in doing the latter it throws up *sprots*, or air-bubbles, which enable the hunters to ascertain where it is, and to spear it. The best time to find it is early in the morning. It may frequently be traced by the dead fish and fish-bones strewed along the banks of the river. The prints, also, of the animal's feet, Page 364 called his *seal*, are of a peculiar formation, and thus it is readily traced. The otter preys during the night, and conceals himself in the daytime under the banks of lakes and rivers, where he generally forms a kind of subterraneous gallery, running for several yards parallel to the water's edge, so that if he should be assailed from one end, he flies to the other. When he takes to the water, it is necessary that those who have otter-spears should watch the bubbles, for he generally vents near them. When the otter is seized, or upon the point of being caught by the hounds, he turns upon his pursuers with the utmost ferocity. Instances are recorded of dogs having been drowned by otters, which they had seized under water, for they can sustain the want of respiration for a much longer time than the dog.

Mr. Daniell, in his "Rural Sports," remarks that hunting the otter was formerly considered as excellent sport, and that hounds were kept solely for that purpose. The sportsmen went on each side of the river, beating the banks and sedges with the dogs. If an otter was not soon found, it was supposed that he had gone to *couch* more inland, and was sought for accordingly. If one was found, the sportsmen viewed his track in the mud, to find which way he had taken.

> "On the soft sand,
> See there his seal impress'd! And on that bank
> Behold the glitt'ring spoils, half-eaten fish,
> Scales, fins, and bones, the leavings of his feast."

Page 365 The spears were used in aid of the dogs. When an otter is wounded, he makes directly to land, where he maintains an obstinate defence:—

"Lo! to yon sedgy bank
He creeps disconsolate; his numerous foes
Surround him, hounds and men. Pierc'd through and through,
On pointed spears they lift him high in air;
Bid the loud horns, in gaily warbling strains,
Proclaim the spoiler's fate: he dies, he dies."

The male otter never makes any complaint when seized by the dogs, or even when transfixed with a spear, but the females emit a very shrill squeal. In the year 1796, near Bridgenorth, on the river Wherfe, four otters were killed. One stood three, another four hours before the dogs, and was scarcely a minute out of sight. In April 1804, the otter-hounds of Mr. Coleman, of Leominster, killed an otter of extraordinary size. It measured from the nose to the end of the tail, four feet ten inches, and weighed thirty-four and a half pounds. This animal was supposed to be eight years old, and to have destroyed for the last five years a ton of fish annually. The destruction of fish by this animal is, indeed, very great, for he will eat none unless it be perfectly fresh, and what he takes himself. By his mode of eating them he causes a still greater consumption, for so soon as an otter catches a fish he drags it on shore, devours it to the vent, and, unless pressed by extreme hunger, always leaves the remainder, and takes to the water in search Page 366 of more. In rivers it is always observed to swim against the stream, in order to meet its prey.

Otters bite very severely, and they will seize upon a dog with the utmost ferocity, and will shake it as a terrier does a rat. The jaws of the otter are so constructed, that even when dead it is difficult to separate them, as they adhere with the utmost tenacity. Otters are frequently found on the banks of the Thames, and a large one was caught in an eel-basket, near Windsor, but the hunting of them is discontinued.

Page 367

THE GREYHOUND.

"Ah! gallant Snowball! what remains,
Up Fordon's banks, o'er Flixton's plains,
Of all thy strength—thy sinewy force,
Which rather flew than ran the course?
Ah! what remains? Save that thy breed
May to their father's fame succeed;
And when the prize appears in view,
May prove that they are Snowballs too."

The perfection to which the greyhound has been brought by persevering care and attention to its breed, distinguishes it alike for beauty, shape, and high spirit, while its habits are mild and gentle in the extreme. These dogs were brought to this great perfection by the late Lord Orford, Major Topham, and others. Snowball,—perhaps one of the best greyhounds that ever ran,—won four cups, couples, and upwards of Page 368 thirty matches, at Malton, and upon the wolds of Yorkshire. In fact, no dog had any chance with him except his own blood. In the November Malton coursing-meeting in 1799, a Scotch greyhound was produced, which had beat every opponent in Scotland. It was then brought to England, and challenged any dog in the kingdom. The challenge was accepted, and Snowball selected for the trial of speed; after a course of two miles, the match (upon which considerable sums were depending) was decided in his favour.

Another dog, which belonged to Sir Henry Bate Dudley, won seventy-four successive matches, without having been once beaten.

Various have been the opinions upon the difference of speed between a well-bred greyhound and a racehorse, if opposed to each other. Wishes had been frequently indulged by the sporting world, that some criterion could be adopted by which the superiority of speed could be fairly ascertained, when the following circumstance accidentally took place, and afforded some information upon what had been previously considered a matter of great uncertainty. In the month of December, some years ago, a match was to have been run over Doncaster race-course for one hundred guineas; but one of the horses having been drawn, a mare started alone, that by running the ground she might ensure the wager, when having run about one mile in the four, she was accompanied by a greyhound bitch, which joined her from the side of the course, Page 369 and emulatively entering into the competition, continued to race with the mare for the other three miles, keeping nearly head and head, and affording an excellent treat to the field by the energetic exertions of each. At passing the distance-post, five to four was betted in favour of the greyhound; when parallel with the stand, it was even betting, and any person might have taken his choice from five to ten: the mare, however, had the advantage by a head at the termination of the course.

The courage and spirit of these dogs is very great. A greyhound ran a hare single-handed and raced her so hard, that, not having time to run through an opening at the bottom of some paling, she and the greyhound made a spring at the same moment at the top of the pales. The dog seized her at the instant she reached it, and in the momentary struggle he slipt between two broken pales, each of which ran into the top of his thighs. In this situation he hung till the horsemen came up, when, to their great surprise, he had the hare fast in his mouth, which was taken from him before he could be released.

I saw a hare coursed on the Brighton Downs some years ago by two celebrated greyhounds. Such was the length of the course, some of it up very steep hills, that the hare fell dead before the dogs, who were so exhausted that they only reached to within six feet of her. This was one of the severest courses ever witnessed.

Page 370 On another occasion, two dogs ran a hare for several miles, and with such speed as to be very soon out of sight of the coursing party. After a considerable search, both the dogs and the hare were found dead within a few yards of each other; nor did it appear that the former had touched the hare. Mr. Daniel, in his "Rural Sports," states that a brace of greyhounds, in Lincolnshire, ran a hare from her seat to where she was killed, a distance, measuring straight, of upwards of four miles, in twelve minutes. During the course there was a good number of turns, which must have very considerably increased the space gone over. The hare ran till she died before the greyhounds touched her.

In the year 1798, a brace of greyhounds, the property of Mr. Courtall of Carlisle, coursed a hare from the Swift, near that city, and killed her at Clemmell, seven miles distant. Both greyhounds were so exhausted, that unless the aid of medical men, who happened to be on the spot, had been immediately given, they would have died, and it was with difficulty they were recovered.

In the year 1818, a black greyhound bitch, the property of Mr. John Heaton, of Scarisbrick, in Lancashire, left her master, forsook the habitation where she had been reared, betook herself to the fields and thickets, and adopted a life of unlimited freedom, defying all the restraints of man. In this state she killed a great number of hares for food, and occa Page 371 sionally made free with the sheep; she, therefore, very soon became a nuisance in the neighbourhood. She had taken her station at the distance of two miles from her master's house, and was generally found near this spot. In consequence of her depredations, many attempts were made to shoot her, but in vain. She eluded, for more than six months, the vigilance of her pursuers. At length she was observed to go into a barn that stood in a field which she frequented. She entered the building through a hole in the wall, and, by means of a rope-snare, was caught as she came out. On entering the barn, three whelps were found about a week old; so that in her savage state she had evidently been visited by a male of her own species. The whelps were (foolishly enough) immediately destroyed. As the bitch herself evinced the utmost ferocity, and, though well secured, vainly attempted to seize every person that approached, she was taken home, and treated with the greatest kindness. By degrees her ferocity abated, and in the course of two months she became perfectly reconciled to her original abode. The following season she ran several courses. There continued a wildness in her look; yet, although at perfect liberty, she did not attempt again to stray away, but seemed quite reconciled to her domestic life.

Few facts can show the high courage of the greyhound more than the following: —

As a gamekeeper of Lord Egremont's was leading a brace of greyhounds in couples, a hare accidentally Page 372 crossed the road in view. This temptation proved so irresistible, that the dogs, by a joint effort, broke suddenly from their conductor, and gave chase, shackled as they were together. When they got up and gave the hare the first turn, it was evidently much to her advantage, as the greyhounds were so embarrassed that it was with great difficulty they could change the direction. Notwithstanding this temporary delay, they sustained no diminution of natural energy, but continued the course through and over various obstructions, till the object of their pursuit fell a victim to their invincible perseverance, after a run of between three and four miles.

In addition to the beauty, elegance, high spirit, and speed of the greyhound, may be mentioned his mild and affectionate disposition, as well as his fidelity and attachment to those who treat him with kindness. They will also show sometimes considerable sagacity, of which the following is an instance:—

Two young gentlemen went to skate, attended only by a greyhound. About the time they were expected home, the dog arrived at the house full speed, and by his great anxiety, by laying hold of the clothes of some of the inmates, and by his significant gestures, he convinced them that something was wrong. They followed the greyhound, and came to the pond. A hat was seen on the ice, near which was a fresh aperture. The bodies of the young gentlemen were soon found, but life was extinct. In this instance the Page 373 sagacity of the dog was extraordinary. Had he possessed the power of speech, he could scarcely have communicated what had taken place more significantly than he did.

I have received the following anecdote from a friend, on whose veracity I can depend:—In the year 1816, a greyhound bitch in pup was sent from the neighbourhood of Edinburgh by a carrier, *viâ* Dumfries, to the neighbourhood of Castle Douglas, in the stewartry of Kirkeudbright. She brought up her litter of pups there, and in the following year was returned by the same route to Edinburgh, from whence she was sent by way of Douglas and Muirkirk to the neighbourhood of Cumnock, in Ayrshire. After remaining there five or six months, she found her way across the country to the house near Castle Douglas where she had brought up her pups. The fact of her crossing the country was ascertained by shepherds, who saw her, accompanied by a pointer-dog. She arrived, accompanied by this dog, who left her almost immediately, and found his way home again. The bitch was bred in East Lothian, and had never been previously either in Ayrshire or Dumfriesshire.

A small Italian greyhound in Bologna, which used at nights to have a kind of jacket put on, to guard him from the cold, went out generally very early in the morning to a neighbouring house, to visit another dog of the same breed which lived there. He always endeavoured, by various coaxing gestures, to prevail Page 374 upon the people of the house to take off his night-jacket, in order that he might play more at ease with his companion. It once happened, when he could not get any one to do him this service, that he found means, by various contortions of his body, rubbing himself against tables and chairs, and working with his limbs, to undress himself without any other assistance. After this trial had succeeded, he continued to practise it for some time, until his master discovered it, who after that undressed him every morning, and let him out of the house. At noon, and in the evening, he always returned home. Sometimes, when he made his morning call, he found the door of the house in which his friend dwelt not yet open. In these cases he placed himself opposite to the house, and by loud barking solicited admittance. But as the noise which he made became troublesome both to the inhabitants of the house and to the neighbours, they not only kept the door shut against him, but endeavoured also to drive him away from the house by throwing stones at him from the windows. He crept, however, so close to the door, that he was perfectly secure against the stones, and now they had to drive him away with a whip. After some time the dog went again to the house, and waited without barking till the door was opened. He was again driven away, upon which he discontinued his visits for a long time. At length, however, he ventured to go once more to the house, and set up a loud barking; placing himself in a situ Page 375 ation where he was both secure against the stones, and could not be seized by the people of the house when they opened the door.

After a considerable time, he one morning saw a boy come to the house, lay hold of the knocker, and strike it against the door, and he observed that upon this process the door was opened. After the boy had been let in, the dog crept along the side of the house to the door, and took his station upon the spot where the boy had stood when he knocked, and where no one who stood close to the door could be seen from within. Here he leaped several times at the knocker, till he raised it and made it strike the door. A person from within immediately called, "Who is there?" but receiving no answer, opened the door, upon which the dog ran in with tokens of great delight, and soon found his way to his friend. Often after this he availed himself of the fortunate discovery which he had made, and his ingenuity was so much admired that it procured him thenceforward free access to his companion's habitation.

While on the subject of greyhounds, I cannot resist the insertion of the following account of one extracted from Froissart:—

When Richard II. was confined in the Castle of Flint, he possessed a greyhound, which was so remarkably attached to him, as not to notice or fawn upon any one else. Froissart says,—"It was informed me Kynge Richard had a grayhounde, called Mathe, Page 376 who always waited upon the kynge, and would know no one else. For whenever the kynge did ryde, he that kept the grayhounde did let him lose, and he wolde streyght runne to the kynge and fawne upon him, and leape with his fore-fete upon the kynge's shoulders. And as the kynge and the Erle of Derby talked togyder in the courte, the grayhounde, who was wont to leape upon the kynge, left the kynge and came to the Erle of Derby, duke of Lancaster, and made to hym the same friendly countenance and chere he was wont to do the kynge. The Duke, who knew not the grayhounde, demanded of the kynge what the grayhounde would do. 'Cosin,' quod the kynge, 'it is a great good token to you, and an evil sygne to me.' 'Sir, how know ye that?' quod the duke. 'I know it well,' quod the kynge: 'the grayhounde maketh you chere this daye as kynge of Englande, as ye shall be, and I shall be deposed; the grayhounde hath this knowledge naturally, therefore take hym to you: he will follow you and forsake me.' The duke understoode well those words, and cheryshed the grayhounde, who would never after followe Kynge Richarde, but followed the Duke of Lancaster." It is not, however, improbable, that the dog thus mentioned was the Irish wolf-dog, as the fact related is more characteristic of that noble animal.

The mild, affable, and serene aspect of the greyhound, constitutes no drawback to its innate sagacity, or grateful attention to its protector, of which the Page 377 unfortunate king Charles I. was so observant, that the remark he made during his troubles is on record, and strictly just as applicable to the instinctive fidelity of the animal. He said the greyhound possessed all the good nature of the spaniel without the fawning.

Washington Irving mentions, that in the course of his reading he had fallen in with the following anecdote, which illustrates in a remarkable manner the devoted attachment of these dogs to their masters: —

"An officer named St. Leger, who was imprisoned in Vincennes (near Paris) during the wars of St. Bartholomew, wished to keep with him a greyhound that he had brought up, and which was much attached to him; but they harshly refused him this innocent pleasure, and sent away the greyhound to his house in the Rue des Lions Saint Paul. The next day the greyhound returned alone to Vincennes, and began to bark under the windows of the tower, where the officer was confined. St. Leger approached, looked through the bars, and was delighted again to see his faithful hound, who began to jump and play a thousand gambols to show her joy. He threw a piece of bread to the animal, who ate it with great good will; and, in spite of the immense wall which separated them, they breakfasted together like two friends. This friendly visit was not the last. Abandoned by his relations, who believed him dead, the unfortunate prisoner received the visits of his greyhound only, during four years' confinement. Whatever weather it Page 378 might be, in spite of rain or snow, the faithful animal did not fail a single day to pay her accustomed visit. Six months after his release from prison St. Leger died. The faithful greyhound would no longer remain in the house; but on the day after the funeral returned to the castle of Vincennes, and it is supposed she was actuated by a motive of gratitude. A jailor of the outer court had always shown great kindness to this dog, which was as handsome as affectionate. Contrary to the custom of people of that class, this man had been touched by her attachment and beauty, so that he facilitated her approach to see her master, and also insured her a safe retreat. Penetrated with gratitude for this service, the greyhound remained the rest of her life near the benevolent jailor. It was remarked, that even while testifying her zeal and gratitude for her second master, one could easily see that her heart was with the first. Like those who, having lost a parent, a brother, or a friend, come from afar to seek consolation by viewing the place which they inhabited, this affectionate animal repaired frequently to the tower where St. Leger had been imprisoned, and would contemplate for hours together the gloomy window from which her dear master had so often smiled to her, and where they had so frequently breakfasted together."

The natural simplicity and peaceable demeanour of the greyhound may have sometimes induced a doubt of its possessing the sagacity, fidelity, and attachment of other dogs; but when he is kindly treated and do Page 379 mesticated, he is capable of showing them to an equal degree with any of the canine race.

Some of the best coursing in England takes place on the Wiltshire Downs, where it is no uncommon sight to see a hare run away from two good dogs without a single turn. Nearly three hundred years ago, Sir Philip Sidney referred to this sport on the Wiltshire Downs in one of his poems, in which he remarks:—

"So, on the downs we see, near Wilton fair,
A hasten'd hare from greedy greyhounds go."

The following account of the Persian greyhound appeared in the "Book of Sports:"—

"The Persian greyhound is much esteemed in its native country, where the nobles, who are excessively fond of the chase, keep a great number of them at a considerable expense, the best and most favoured dogs frequently having their collars and housings covered with precious stones and embroidery.

"These greyhounds are employed in coursing hares in the plains, and in chasing the antelope. As the speed of the antelope is greater than that of the greyhound, the Persians train hawks for the purpose of assisting the dog in this kind of chase. The hawks when young are fed upon the head of a stuffed antelope, and thus taught to fly at that part of the animal. When the antelope is discovered, the hawk is cast off, which, fastening its talons in the animal's head, im Page 380 pedes its progress, and thus enables the greyhounds to overtake it. The chase, however, in which the Persians chiefly delight, and for which those greyhounds are most highly valued, is that of the ghoo-khur, or wild ass. This animal, which generally inhabits the mountainous districts, is extremely shy, and of great endurance, and is considered by the Persians as one of the swiftest of all quadrupeds. These qualities, and the nature of the ground over which it is usually chased, render the capture of the wild ass very uncertain, and its pursuit extremely hazardous to the sportsman.

"When the Persians go out to hunt the wild ass, relays of greyhounds are placed at various distances in the surrounding country, in such directions as are most likely to be traversed by the object of pursuit; so that when one relay is tired, there is another fresh to continue the chase. Such, however, is the speed and endurance of the ghoo-khur, that it is seldom fairly run down by the greyhounds; its death generally being achieved by the rifle of some lucky horseman. The Persians evince great skill and courage in this arduous sport; riding, rifle in hand, up and down precipitous hills, over stony paths, and across ravines and mountain streams, which might well daunt our boldest turf-skimming Meltonians.

"Though several Persian greyhounds have at different times been brought to this country, the breed can scarcely be considered as established here. The specimen, however, (a female), from which Mr. Hamil Page 381 ton painted the picture from which our engraving is taken, was bred in this country. She was then supposed to be the only Persian greyhound bitch in England."

The Persian greyhound is very handsome. "One of the finest species of dog I have ever seen," says an interesting writer, "is a sort of greyhound which the Persians rear to assist them in the chase. They have generally long silken hair upon their quarters, shoulders, ears, and tail; and I think them as handsome, and considerably more powerful and sagacious, than our own greyhounds. I have sometimes seen a spirited horse break loose, and run away at full speed, when one of these dogs has set after him like an arrow, and soon getting ahead of him, taken an opportunity of seizing the bridle in his teeth, which he held so firmly, that though he was not strong enough to stop the horse, yet, as he was dragged along, he continued to pull and confine the horse, so as to impede him very much, till some person was able to overtake and secure him."

Col. Hutchinson says, that "In Persia and many parts of the East greyhounds are taught to assist the falcon in the capture of deer. When brought within good view of a herd the bird is flown, and at the same moment the dog is slipped. The rapid sweep of the falcon soon carries him far in advance. It is the falcon who makes the selection of the intended victim—which appears to be a matter of chance—and a properly-trained greyhound will give chase to none other, Page 382 however temptingly close the alarmed animals may pass him. The falcon is instructed to aim at the head only of the gazelle, who soon becomes bewildered; sometimes receiving considerable injury from the quick stroke of its daring adversary. Before long the gazelle is overtaken by the greyhound. It is not always easy to teach a dog to avoid injuring the bird, which is so intent upon its prey as utterly to disregard the approach of the hound. Death would probably be the penalty adjudged to him for so heinous an offence; for a well-trained falcon is of great value. You can readily imagine that neither it nor the greyhound could be properly broken unless the instructor possessed much judgment and perseverance. The sport is very exciting; but the spectator must be well-mounted, and ride boldly, who would closely watch the swift, varying evolutions of the assailing party, and the sudden evasions of the helpless defendant."

Page 383

THE POINTER.

"The subtle dog scours with sagacious nose
Along the field, and snuffs each breeze that blows;
Against the wind he takes his prudent way,
While the strong gale directs him to the prey.
Now the warm scent assures the covey near;

He treads with caution, and he points with fear.
The fluttering coveys from the stubble rise,
And on swift wing divide the sounding skies;
The scatt'ring lead pursues the certain sight,
And death in thunder overtakes their flight." —Gay.

Page 384 This dog has been crossed and re-crossed so often with the fox-hound, the setter, and the old Spanish pointer, that the originality of the present breed may be questioned, especially as the pointer has been less noticed by writers on dogs than any other of the species. How well do I recollect in my early youth seeing the slow, heavy, solemn-looking, and thick-shouldered Spanish pointer, tired with two or three hours' work in turnips, and so stiff after it the next day, as to be little capable of resuming his labours. And yet this dog, fifty years ago, was to be met with all through England. How different is the breed at the present time! By crossing with the fox-hound, they have acquired wonderful speed, and a power of endurance equally surprising, while their shape is beautiful and their sense and animation strongly marked in their intelligent countenances.

The old pointers were either nearly white or variegated with large liver-coloured patches. We now see them either completely liver-coloured, or of a flea-bitten blue or grey, or else black, with fine sterns showing much blood, and extremely thin ears. There can be Page 385 no doubt but that the crosses by which they have obtained the qualities and appearance I have mentioned, render the task of breaking them in to point, back, and drop to charge, one of no small difficulty. These habits, having been acquired in the original breed, had probably become hereditary; but the mixture with dogs which had not these inherent qualities, has introduced volatility and impatience not easily to be overcome. It is also a fact, that if a pointer, notwithstanding this disposition, should at last become perfectly well broke in, or, as it is called, highly broke, he loses much of his natural sagacity. His powers of endurance are, however, very great. A friend of mine, an ardent sportsman, had a pointer crossed with a foxhound, and it was the only one he had. Day after day he took this dog out with him, from day-break till late in the evening, and he never flagged or showed fatigue. It was calculated that he could not traverse less than one hundred and twenty miles each day. This dog showed extraordinary sagacity. While hunting in a large fallow field he made a point, and then slowly and cautiously proceeded, closely followed by his master. In this way he led him over a good part of the field, till it was supposed the dog was drawing on the scent of a hare, which had stolen away. At last he set off running as hard as he could, made a large circuit to the left, and then came to a point immediately opposite to his master, who then advanced and put up a covey of birds between him and the dog.

Page 386 The following is a proof of the perfection to which pointers may be brought. The friend above referred to went out shooting with a gentleman celebrated for the goodness of his breed. They took the field with eight of these dogs. If one pointed, all the rest immediately backed steadily. If a partridge was shot, they all dropped to charge, and whichsoever dog was called to bring the bird, the rest never stirred till they were told to do so. Dogs thus broke in are of great value, and bring large prices; from fifty to a hundred guineas have been given for a good dog.

Pointers frequently show extraordinary sense, especially in their own peculiar vocation. Thus a pointer has been known to refuse to hunt for a person who had previously missed every bird the dog had found. He left him with every mark of disgust, nor could any coaxing induce him to continue with his unsportsman-like companion.

Three pointers were taken out grouse-shooting in Ireland. They were all of the same breed, or rather nearly related to each other, one being the grandmother, the other her daughter, and the third her granddaughter. The latter, who could get over the ground quicker than the others, put up first one pack of grouse, and then another, for which faults she was flogged again and again. Having done the same thing the third time, the steady old grandmother was so provoked, that she ran at the culprit, knocked her over and over, and did not cease to attack her till she had driven her home. Page 387 The authenticity of this anecdote need not be doubted. It is a proof of the extraordinary sense of a dog, and is corroborated by a fact already mentioned in the introductory remarks (p. 33), of one dog attacking another for having misconducted himself.

Some very bad shots went out partridge-shooting, attended by a very good, old, steady pointer. After shooting for some hours with very little success, they began to amuse themselves by firing at a piece of paper stuck on a post. The disgust of the old dog at this proceeding may be imagined—he ran home.

In further proof of the dislike a pointer will show to a bad shot, I will adduce the following anecdote mentioned by Captain Brown. A gentleman, on his requesting the loan of a pointer-dog from a friend, was informed by him that the dog would behave very well so long as he could kill his birds; but if he frequently missed them, it would run home and leave him. The dog was sent, and the following day was fixed for trial; but, unfortunately, his new master was a remarkably bad shot. Bird after bird rose and was fired at, but still pursued its flight untouched, till, at last, the pointer became careless, and often missed his game. As if seemingly willing, however, to give one chance more, he made a dead stop at a fern-bush, with his nose pointed downward, the fore-foot bent, and his tail straight and steady. In this position he remained firm till the sportsman was close to him, with both barrels cocked, then moving steadily forward Page 388 for a few paces, he at last stood still near a bunch of heather, the tail expressing the anxiety of the mind by moving regularly backwards and forwards. At last out sprung a fine old blackcock. Bang, bang, went both barrels, but the bird escaped unhurt. The patience of the dog was now quite exhausted; and, instead of dropping to charge, he turned boldly round, placed his tail between his legs, gave one howl, long and loud, and set off as fast as he could to his own home.

I have seen a pointer leap on the top of a high gate, in going from one field to another, and remain steadily there till I came up to him. He had suddenly come on the scent of birds, and made his point from his uncomfortable situation on the gate. Captain Brown also relates a nearly similar instance of the stanchness of a pointer, which he received from a friend of his. This gentleman was shooting in Scotland, when one of his dogs, in going over a stone wall, about four feet high, got the scent of some birds on the other side of the wall, just as she made the leap. She hung by her fore-legs, appearing at a distance as if they had got fastened among the stones, and that she could not extricate herself. In this position she remained until her master came up. It was then evident that it was her caution for fear of flushing some birds on the other side of the wall, which prevented her from taking the leap, or rather, which was the cause of her making this extraordinary point.

Page 389 Mr. Daniel, in his "Rural Sports," mentions the circumstance of two pointers having stood at one point an hour and a quarter, while an artist took a sketch of them.

A dog of the pointer kind, brought from South Carolina in an English merchant vessel, was a remarkable prognosticator of bad weather. Whenever he was observed to prick up his ears, scratch the deck, and rear himself to look to the windward, whence he would eagerly snuff up the wind, if it was then the finest weather imaginable, the crew were sure of a tempest succeeding; and the dog became so useful, that whenever they perceived the fit upon him, they immediately reefed the sails, and took in their spare canvas, to prepare for the worst. Other animals are prognosticators of weather also; and there is seldom a storm at sea, but it is foretold by some of the natural marine barometers on board, many hours before the gale.

The following circumstance serves also to prove the extreme stanchness of a pointer. It is related by Captain Brown:—

"A servant who used to shoot for Mr. Clutterbuck of Bradford, had, on one occasion, a pointer of this gentleman's, which afforded him an excellent day's sport. On returning, the night being dark, he dropped, by some chance, two or three birds out of his bag, and on coming home he missed them. Having informed a fellow-servant of his loss, he requested him to get up early the next morning, and seek for them near the Page 390 turnpike, being certain that he had brought them as far as that place. The man accordingly went there, and not a hundred yards from the spot mentioned by his companion, he, to his surprise, found the pointer lying near the birds, and where he probably had remained all night, although the poor animal had been severely hunted the day before."

For the following instance of the sagacity of a pointer, I am indebted to Lord Stowell. Mr. Edward Cook, after having lived some time with his brother at Tugsten, in Northumberland, went to America, and took with him a pointer-dog, which he lost soon afterwards, while shooting in the woods near Baltimore. Some time after, Mr. and Mrs. Cook, who continued to reside at Tugsten, were alarmed at hearing a dog in the night. They admitted it into the house, and found that it was the same their brother had taken with him to America. The dog lived with them until his master returned home, when they mutually recognised each other. Mr. Cook was never able to trace by what vessel the dog had left America, or in what part of England it had been landed. This anecdote confirms others which I have already mentioned relative to dogs finding their way back to this country from considerable distances.

Lieutenant Shipp, in his Memoirs, mentions the case of a soldier in India, who, having presented his dog to an acquaintance, by whom he was taken a distance of four hundred miles, was surprised to see Page 391 him back in a few days afterwards. When the faithful animal returned, he searched through the whole barracks for his master, and at length finding him asleep, he awoke him by licking his face.

Pointers have been known to go out by themselves for the purpose of finding game, and when they have succeeded, have returned to their master, and by significant signs and gestures have led them directly to the spot.

The mental faculties of pointers are extremely acute. When once they become conscious of their own powers, and of what is required of them, they seldom commit a fault, and do their duty with alacrity and devotion. Old pointers are apt to hunt the hedgerows of a field before they begin to quarter the ground. I have seen dogs severely rated and punished for doing this, but the cause is obvious. They are aware that game is more frequently to be found in hedgerows than in the open ground, and therefore very naturally take the readiest way of finding it.

An interesting exhibition of clever dogs took place in London in the summer of 1843, under the auspices of M. Léonard, a French gentleman of scientific attainments and enlightened character, who had for some years directed his attention to the reasoning powers of animals, and their cultivation. Two pointers, Braque and Philax, had been the especial objects of his instruction, and their intellectual capacities had been excited in an extraordinary degree. A writer in the "Atlas" newspaper thus speaks of the exhibition of Page 392 these animals:—M. Léonard's dogs are not merely clever, well-taught animals, which, by dint of practice, can pick up a particular letter, or can, by a sort of instinct, indicate a number which may be asked for; they call into action powers which, if not strictly intellectual, approximate very closely to reason. For instance, they exert memory. Four pieces of paper were placed upon the floor, which the company numbered indiscriminately, 2, 4, 6, 8. The numbers were named but once, and yet the dogs were able to pick up any one of them at command, although they were not placed in regular order. The numbers were then changed, with a similar result. Again, different objects were placed upon the floor, and when a similar thing—say a glove—was exhibited, one or other of the animals picked it up immediately. The dogs distinguish colours, and, in short, appear to understand everything that is said to them.

The dog Braque plays a game of dominoes with any one likes. We are aware that this has been done before; but when it is considered that it is necessary to distinguish the number of spots, it must be admitted that this requires the exercise of a power little inferior to reason. The dog sits on the chair with the dominoes before him, and when his adversary plays, he scans each of his dominoes with an air of attention and gravity which is perfectly marvellous. When he could not match the domino played, he became restless and shook his head, and gave other indications of his Page 393 inability to do so. No human being could have paid more attention. The dog seemed to watch the game with deep interest, and what is more, he won.

Another point strongly indicative of the close approach to the reasoning powers, was the exactness with which the dogs obeyed an understood signal. It was agreed that when three blows were struck upon a chair, Philax should do what was requested; and when five were given, that the task should devolve on Braque. This arrangement was strictly adhered to. We do not intend to follow the various proofs which were afforded of the intelligence of the dogs; it is sufficient to say that a multiplicity of directions given to them were obeyed implicitly, and that they appeared to understand what their master said as well as any individual in the room.

M. Léonard entered into a highly-interesting explanation of his theory regarding the intellectual powers of animals, and the mode he adopts to train and subdue horses, exhibiting the defects of the system generally pursued. His principle is, that horses are not vicious by nature, but because they have been badly taught, and that, as with children, these defects may be corrected by proper teaching. M. Léonard does not enter into these inquiries for profit, but solely with a scientific and humane view, being desirous of investigating the extent of the reasoning powers of animals.

It does not appear possible that dogs should be Page 394 educated to the extent of those of M. Léonard, unless we can suppose that they acquire a tolerably exact knowledge of language. That they in reality learn to know the meaning of certain words, not merely when addressed to them, but when spoken in ordinary conversation, is beyond a doubt; although the accompanying looks and movements in all likelihood help them in their interpretation. We have known a small spaniel, for instance, which thoroughly understood the meaning of "out," or "going out," when spoken in the most casual way in conversation. A lady of our acquaintance has a dog which lives at enmity with another dog in the neighbourhood, called York, and angrily barks when the word York is pronounced in his hearing.

A well-known angler was in the habit of being attended by a pointer-dog, who saved him the trouble of a landing-net in his trout-fishing excursions. When he had hooked a fish and brought it near the bank, the dog would be in readiness, and taking the fish behind the head, would bring it out to his master.

A writer, who endeavours to prove the existence of souls analogous to the human in animals, relates the following remarkable fact, of which he was himself an eye-witness. He says: —

"I was with a gentleman who resides in the country, in his study, when a pointer-dog belonging to him came running to the door of the room, which was shut, scratching and barking till he was Page 395 admitted. He then used supplicating gestures of every kind, running from his master to the stair behind which his gun stood, then again to his master, and back to the gun. The gentleman now comprehended something of his dog's meaning, and took up his gun. The dog immediately gave a bark of joy, ran out at the door, returned, and then ran to the back-door of the house, from whence he took the road to a neighbouring hill.

"His master and I followed him. The dog ran, highly pleased, a little distance before us, showing us the way we should take. After we had proceeded about forty paces, he gave us to understand that we should turn to the left, by pressing repeatedly against his master, and pushing him towards the road that turned to the left. We followed his direction, and he accompanied us a few paces, but suddenly he turned to the right, running round the whole of the hill. We still proceeded to the left, slowly up the ascent, till we were nearly arrived at its summit, the dog in the meantime making the circuit of the hill to the right. He was now already higher than we were, when he gave a sudden bark, and that moment a hare ran before the muzzle of his master's gun, and, of course, met her fate."

A gentleman had a pointer so fleet that he often backed him to find birds in a ten-acre field within two minutes, if there were birds in it. On entering the field, he seemed to know by instinct where the Page 396 birds would lie, generally going up to them at once. His nose was so good, that with a brisk wind, he would find his game a hundred and fifty yards off across the furrows. He could tell whether a bird was hit, and if so would retrieve it some fields off from where it was shot. He would never follow a hare unless it was wounded. He would point water-fowl as well as all birds of game, and has been seen pointing a duck or a moor-hen with the water running over his back at the time. Nothing seemed to spoil this dog, not even rat and otter hunting, in both of which he was an adept, as he knew his business; and although he would rattle through a wood, he was perfectly steady the next minute out of cover. He has been known to continue at a point two hours. In high turnips he would contrive to show his master where he was, standing sometimes on his hind legs only, so that his head and fore-quarters might be seen. On one occasion he came at full speed so suddenly on a hare, that he slipped up, and fell nearly on his back. In this position he did not move, and it was thought he was in a fit, till the hare jumped up and was killed, when the dog righted himself. So steady was he in backing another dog when game was found, that he once caught sight of a point at the moment of jumping a stile, and balanced himself on it for several seconds till he fell. Once when hunting with a young pointer, who had only been taken into the field two or three times, in order to show him some Page 397 birds before the shooting season, the following occurrence took place. The old dog found some birds in the middle of the field, and pointed them steadily. The puppy had been jumping and gambolling about, with no great hunt in him, and upon seeing the old dog stand, ran playfully up to him. He was, however, seized by the neck, and received a good shaking, which sent him away howling, and his companion then turned round and steadied himself on his point, without moving scarcely a yard. This anecdote is extracted from Hone's "Year Book," and the writer of it goes on to say,—"What dog is there possessing the singular self-denial of the pointer or setter? The hound gives full play to his feelings; chases, and babbles, and kicks up as much riot as he likes, provided he is true to his game; the spaniel has no restraint, except being kept within gun-shot; the greyhound has it all his own way as soon as he is loosed; and the terrier watches at a rat's hole, because he cannot get into it: but the pointer, at the moment that other dogs satisfy themselves, and rush upon their game, suddenly stops, and points with almost breathless anxiety to that which we might naturally suppose he would eagerly seize. The birds seen, the dog creeps after them cautiously, stopping at intervals, lest by a sudden movement he should spring them too soon. And then let us observe and admire his delight when his anxiety—for it is anxiety—is crowned with success—when the bird falls, and he lays it joyfully at his Page 398 master's feet. A pointer should never be ill-used. He is too much like one of us. He has more headpiece than all the rest of the dogs put together. Narrowly watch a steady pointer on his game, and see how he holds his breath. It is evident he must stand in a certain degree of pain, for we all know how quickly a dog respires. And when he comes up to you in the field he puffs and blows, and his tongue is invariably hanging out of his mouth. We never see this on a point, and to check it suddenly must give the dog pain. And yet, how silent he is! how eager he looks! and if a sudden hysteric gasp is heard, it ceases in a moment. Surely he is the most perfect artist of the canine race."

Some of my readers may like to know that the best breaker of pointers I have yet met with is Mr. Lucas, one of the keepers of

Richmond Park. He perfectly understands his business, and turns out his pointers in a way which few can equal.

In August 1857, a gentleman residing at Ludlow, in Shropshire, had a pointer bitch, which produced seven puppies. Six of them were drowned, and one left. On the servant going the next morning to give her some milk, she found, besides the puppy, a hedgehog, which had been in the garden some years, most comfortably curled up with them. She took it away, but my informant being told that it had got back again, he went to see it. The pointer was licking it, and appeared quite as fond of it as of her own puppy. Page 399 He again had it removed, the bitch following, and whining with evident anxiety to have it restored to her. This was the more remarkable, as on previous occasions she had tried to kill the hedgehog. This strange affection can only be accounted for by an abundant flow of milk, which distended and hurt her, occasioned by her other puppies having been destroyed, and she, therefore, seized on the hedgehog to relieve her, however incongruous it might be to her former feelings towards it.

Page 400

THE SETTER.

The old English setter (says Capt. Brown), was originally derived from a cross between the Spanish Page 401 pointer and the large water-spaniel, and was justly celebrated for his fine scent. It is difficult now to say what a setter really is, as the original breed has been crossed with springers, stag and blood-hounds. The Irish breed of setters is considered better than either the English or Scotch, and a fine brace has been frequently known to fetch fifty guineas. Youatt says that the setter is evidently the large spaniel improved in size and beauty, and taught to mark his game by setting or crouching. He is more active than the pointer, but has not so much patient steadiness. It is extremely difficult to decide between the merits of the setter and pointer as dogs for shooting over. Some authors prefer one, some the other. "Craven" says, that in his opinion Russian setters are better than English, in nose, sagacity, and every other qualification that a dog ought to possess.

Col. Hutchinson relates that he was "partridge-shooting the season before last with an intimate friend. The air was soft, and there was a good breeze. We came upon a large turnip-field, deeply trenched on account of its damp situation. A white setter, that habitually carried a lofty head, drew for awhile, and then came to a point. We got up to her. She led us across some ridges, when her companion, a jealous dog (a pointer), which had at first backed correctly, most improperly pushed on in front, but, not being able to acknowledge the scent, went off, clearly imagining the Page 402 bitch was in error. She, however, held on, and in beautiful style brought us up direct to a covey. My friend and I agreed that she must have been but little, if at all, less than one hundred yards off when she first winded the birds; and it was clear to us that they could not have been running, for the breeze came directly across the furrows, and she had led us in the wind's eye. We thought the point the more remarkable, as it is generally supposed that the strong smell of turnips diminishes a dog's power of scenting birds."

The same able author says, that on one occasion when a near relation of his was shooting on the banks of the Forth, he killed a partridge that was flying across the river. As he had no retriever with him, he almost regretted having fired; but, to his surprise, his setter, Dove, jumped into the river, although she had never previously (to the writer's knowledge), attempted to swim, seized it, and deposited it safely on the bank. She never had retrieved before, and was not particularly good at "seeking dead."

"During my residence in the country," says M. Huet, "I had a gamekeeper who was very skilful in the art of training dogs.

Among others of various kinds which he trained was a large old English setter, with which he had succeeded so well that he could use him both for hunting and shooting.

"This dog did always as much as could be done by any of his race, in whatever kind of sport he was em Page 403 ployed; he even invented advantageous manœuvres himself, which the gamekeeper affirmed he had never taught him.

"Once, after I had been already several hours returned from hunting with my people, the dog came running across the yard with a hare upon his back, which he held by the ear, so as to carry her in the most convenient manner to the kitchen from the considerable distance where he must have killed her.

"Upon another occasion he showed an extraordinary degree of judgment and fidelity. The gamekeeper had, on one of the short days of December, shot at and wounded a deer. Hoping to run him down before night, he instantly put the dog upon the track, which followed it at full speed, and soon was out of sight. At length it grew dark, and the gamekeeper returned home, thinking he should find the setter arrived there before him; but he was disappointed, and became apprehensive that his dog might have lost himself, or fallen a prey to some ravenous animal. The next morning, however, we were all greatly rejoiced to see him come running into the yard, whence he directly hastened to the door of my apartment, and, on being admitted, ran, with gestures expressive of solicitude and eagerness, to a corner of the room where guns were placed. We understood the hint, and, taking the guns, followed him. He led us not by the road which he himself had taken out of the wood, but by beaten paths half round it, and then by several wood-cutters' Page 404 tracks in different directions, to a thicket, where, following him a few paces, we found the deer which he had killed. The dog seems to have rightly judged that we should have been obliged to make our way with much difficulty through almost the whole length of the wood, in order to come to the deer in a straight direction, and he therefore led us a circuitous but open and convenient road. Between the legs of the deer, which he had guarded during the night against the beasts of prey that might otherwise have seized upon it, he had scratched a hole in the snow, and filled it with dry leaves for his bed. The extraordinary sagacity which he had displayed upon this occasion rendered him doubly valuable to us, and it therefore caused us very serious regret when, in the ensuing summer, the poor animal went mad, possibly in consequence of his exposure to the severe frost of that night, and it became necessary for the game-keeper to shoot him, which he could not do without shedding tears. He said he would willingly have given his best cow to save him; and I confess myself that I would not have hesitated to part with my best horse upon the same terms."

Mr. Torry, of Edinburgh, had a setter bitch which possessed great powers, and especially in finding lost articles, as she would, whenever she was desired, go in search of anything. On one occasion his servant lost a favourite whip in the middle of a moor, and he did not discover or make known this loss till they were Page 405 about a mile distant from the spot where it was dropped. Mr. Torry ordered the servant to go back and bring it, as he stated he was quite certain of the spot where he had dropped it; but after searching for nearly an hour, the servant returned and said he could not recover it, upon which Mr. Torry told his setter to go back for the whip. She started off instantly, and in less than five minutes the lost article was at his feet.

The same dog did a great many other curious things: she would ring the bell, fetch her master's slippers, or bring his youngest son, when required to do so, from another room; which last she effected by taking hold of his pinafore with her mouth, and running before him sideways to his master's chair.

A large setter, ill with the distemper, had been most tenderly nursed by a lady for three weeks. At length he became so weak as to be placed on a bed, where he remained three days in a dying situation. After a short absence, the lady, on re-entering the room, observed him to fix his eyes attentively on her, and make an effort to crawl across the bed towards her. This he accomplished evidently for the sole purpose of licking her hands, which, having done, he expired without a groan. "I am," says Mr. Blaine, "as convinced that the animal was sensible of his approaching dissolution, and that this was a last forcible effort to express his gratitude for the care taken of him, as I am of my Page 406 own existence; and had I witnessed this proof of excellence alone, I should think a life devoted to the amelioration of the condition of dogs far too little for their deserts."

There is a curious and interesting anecdote related of a setter who had formed a great friendship with a cat. They were, in fact, inseparable companions, and evidently had a great love for each other. As a sporting dog the setter had few equals, but he constantly showed his disgust when obliged to accompany a bad shot into the fields. After one of the shooting seasons was over, his master took a house in London, and carried his setter with him, who was seated with the footman on the box of the carriage. It appears that the dog had not forgotten his favourite, the cat, for he disappeared from the house, and was absent for some days. He at length returned to his master's house in the country, and brought back the cat with him. How he contrived to find his way backwards and forward, and how he persuaded the cat to accompany him, are mysteries which it would be useless to attempt to solve. The fact, however, would seem to be satisfactorily vouched for.

Setters are known to be subject to strange freaks. A gentleman had one which he had shot to for three years. Upon one occasion he took the dog out, and fired seven or eight times at birds the dog had found him; but having missed them all, the animal returned Page 407 home, evidently disgusted. In the evening his owner took him out again and killed every shot, which procured a reconciliation between the dog and its master.

The late Dr. Hugh Smith related the following circumstance of a setter dog, and maintained that a bitch and a dog may fall passionately in love with each other. As the doctor was travelling from Midhurst into Hampshire, the dogs, as usual in country places, ran out barking as he was passing through a village; and amongst them he observed a little ugly mongrel, that was particularly eager to ingratiate himself with a setter bitch that accompanied him. Whilst stopping to water his horse, he remarked how amorous the mongrel continued, and how courteous the setter seemed to her admirer. Provoked to see a creature of Dido's high blood so obsequious to such mean addresses, the doctor drew one of his pistols and shot the dog; he then had the bitch carried on horseback for several miles. From that day, however, she lost her appetite, ate little or nothing, had no inclination to go abroad with her master, or attend to his call, but seemed to repine like a creature in love, and express sensible concern for the loss of her gallant. Partridge season came, but Dido had no nose. Some time after she was coupled to a setter of great excellence, which with no small difficulty had been procured to get a breed from, and all the caution which even the doctor himself could take was strictly exerted, that the whelps might be pure and unmixed; yet not a puppy did Dido Page 408 bring forth but what was the picture and colour of the mongrel that he had so many months before destroyed. The doctor fumed, and, had he not personally paid such attention to preserve the intercourse uncontaminated, would have suspected that some negligence had occasioned this disappointment; but his views were in many subsequent litters also defeated, for Dido never produced a whelp which was not exactly similar to the unfortunate dog which was her first and murdered lover.

This anecdote may appear strange or untrue to some people; but it is an undoubted fact, and in some degree corroborates Dr. Smith's account that the late Sir Gore Ouseley had a Persian mare which produced her first foal by a zebra in Scotland. She was afterwards a brood-mare in England, and had several foals, every one of which had the zebra's stripes on it. That the force of imagination influences some brutes cannot be doubted. A gentleman had a small spaniel which had one of her legs broken when pregnant. When she littered, one of the whelps had one of her hind legs broken—the limb was contracted—a perfect callus formed, in everything resembling the leg of the dam.

Setters are difficult to break; but when well broken are invaluable as sporting dogs, for they will work all day if they can occasionally find water. John Dudley, duke of Northumberland, is said to have been the first that broke a setter dog to the net, about the year 1555.

Page 409 Col. Hutchinson says that a French lady, who is fond of animals, at his request committed the following anecdote to paper:—

"My dear Médor, a beautiful red-and-white setter, was remarkable, I am told, for many rare qualities as a sporting dog; but, of course, none of these could be compared, in *my* eyes, to his faithfulness and sagacity. I looked upon him as a friend; and I know that our affection was mutual. I could mention several instances of his intelligence—I might say, reflection; but one in particular gave me such delight that, though years have since passed away, all the circumstances are as fresh in my memory as if they had occurred but yesterday. I was returning from school at Versailles; and having rang uselessly for a little time at the front door, I went round to the carriage-gate to have a chat with my silky-haired favourite. He barked anxiously, thrust his cold nose through an opening near the ground, scratched vigorously to increase its size, and in numerous ways testified great joy at again hearing my voice. I put my hand under the gate to caress him; and while he was licking it, I said in jest, but in a distinct, loud voice, 'Dear Médor, I am shut out—go, bring me the keys.' It so happened that the stable where they usually hung was not closed. Médor ran off, and in a few seconds returned and placed them in my hands. I will not attempt to describe *my* gratification at such a striking proof of his intelligence, nor *his* evident pride at seeing me enter Page 410 the hall, nor yet the fright of the servant at thinking how long the street-door must have been carelessly left open. 'Médor deserves that his life should be written,' said I to my uncle, when afterwards telling him the whole story; 'I am sure his deeds are as wonderful as those related of the 'Chiens célèbres' by De Fréville.'

"My setter was immediately declared 'Keeper of the Keys,' and forthwith invested with all the rights of office. Nor was this confidence misplaced. He would never give up his charge to any one but to my uncle or myself; and always seemed fully sensible of the dignity and responsibility of his new position."

Tolfrey gives, in his "Sportsman in France," so beautiful an instance of a setter's untutored intelligence leading him to see the advantage of placing running birds between himself and the gun, that I will relate it.

"On gaining some high ground, the dog drew and stood. She was walked up to, but to my astonishment we found no birds. She was encouraged, and with great difficulty coaxed off her point. She kept drawing on, but with the same ill success.

"I must confess I was for the moment sorely puzzled; but knowing the excellence of the animal, I let her alone. She kept drawing on for nearly a hundred yards—still no birds. At last, of her own accord, and with a degree of instinct amounting almost to the faculty of reason, she broke from her point, and Page 411 dashing off to the right made a *détour*, and was presently straight before me, some three hundred yards off, setting the game whatever it might be, as much as to say, 'I'll be —— if you escape me this time.' We walked steadily on; and when within about thirty yards of her, up got a covey of red-legged partridges, and we had the good fortune to kill a brace each.

"It is one of the characteristics of these birds to run for an amazing distance before they take wing; but the sagacity of my faithful dog baffled all their efforts to escape. We fell in with several coveys of these birds during the day, and my dog ever after gave them the double, and kept them between the gun and herself."

Page 412

THE COMFORTER, OR LAP-DOG PUG.

THE PUG DOG.

"My pug makes a bad pet; he is useless in the field, is somewhat snappish, has little sagacity, and is very cowardly: but there is an air of *bon ton* about him which renders him a fashionable appendage to a fine lady." —*Parisian Gossip.*

Pugs came into fashion, and probably first into this country, in the early part of the reign of William the Third, and were then called Dutch pugs. At that time they were generally decorated with orange rib Page 413 bons, and were in great request amongst the courtiers, from the king being very partial to them.

It is difficult to say how this partiality arose, though it may perhaps be accounted for by the following anecdote, related in a scarce old book, called "Sir Roger Williams' Actions in the Low Countries," printed in 1618.

"The Prince of Orange (father of William III.) being retired into the camp, Julian Romero, with earnest persuasions, procured license of the Duke D'Alva to hazard a *camisado*, or night attack, upon the prince. At midnight Julian sallied out of the trenches with a thousand armed men, mostly pikes, who forced all the guards that they found in their way into the place of arms before the Prince's tent, and killed two of his secretaries. The Prince himself escaped very narrowly, for I have often heard him say that he thought but for a dog he should have been taken or slain. The attack was made with such resolution, that the guards took no alarm until their fellows were running to the place of arms, with their enemies at their heels, when this dog, hearing a great noise, fell to scratching and crying, and awakened him before any of his men; and though the Prince slept armed, with a lacquey always holding one of his horses ready bridled and saddled, yet, at the going out of his tent, with much ado he recovered his horse before the enemy arrived. Nevertheless, one of his equerries was slain taking Page 414 horse presently after him, as were divers of his servants. The Prince, to show his gratitude, until his dying day kept one of that dog's race, and so did many of his friends and followers. These animals were not remarkable for their beauty, being little white dogs, with crooked noses, called *Camuses* (flat-nosed)."

It is difficult to account for the origin of this breed of dogs. So far from having any of the courage of the bulldog, which they resemble somewhat in miniature, they are extremely cowardly. They are also occasionally treacherous in their disposition, and will take strong dislikes to particular persons.

The passion of the late Lady Penrhyn for pugs was well known. Two of these, a mother and daughter, were in the eating-room of Penrhyn Castle during the morning call of a lady, who partook of luncheon. On bonnets and shawls being ordered for the purpose of taking a walk in the grounds, the oldest dog jumped on a chair, and looked first at a cold fowl, and then at her daughter. The lady remarked to Lady Penrhyn that they certainly had a design on the tray. The bell was therefore rung, and a servant ordered to take it away. The instant the tray disappeared, the elder pug, who had previously played the agreeable with all her might to the visitor, snarled and flew at her, and during the whole walk followed her, growling and snapping at her heels whenever opportunity served. Page 415 The dog certainly went through two or three links of inference, from the disappearance of the coveted spoil to Lady Penrhyn's order, and from Lady Penrhyn's order to the remark made by her visitor.

Monsieur Blaze, in his "History of Dogs," mentions one who was taught to pronounce several words. The editor of the "Dumfries Courier" has declared most solemnly that he "heard a pug repeatedly pronounce the word 'William,' almost as distinctly as ever it was enunciated by the human voice. He saw the dog lying on a rug before the fire, when one of his master's sons, whose name is William, and to whom he is more obedient than to any one else, happened to give him a shove, when the animal ejaculated, for the first time, the word 'William.' The whole party were as much amazed as Balsam was when his ass spoke; and though they could hardly believe their own ears, one of them exclaimed, 'Could you really find it in your heart to hurt the poor dog after he has so distinctly pronounced your name?' This led to a series of experiments, which have been repeated for the satisfaction of various persons, but still the animal performs with difficulty. When his master seizes his fore-legs, and commands him to say 'William,' he treats the hearer With a gurring voluntary; and after this species of music has been protracted for a longer or a shorter period, his voice seems to fall a full octave before he comes out with the important word."

Page 416 In the "Bibliothèque Germanique," published in 1720, there is an account of a dog at Berlin, who was made to pronounce a few words, but the one which he ejaculated most distinctly was "Elizabeth." Sir William Gell also had a dog which was well known to repeat some words, but it should be mentioned that he never did this except his master held his jaws in a peculiar way. [R]

It has been said of the pug dog that he is applicable to no sport, appropriated to no useful purpose, susceptible of no predominant passion, and in no way remarkable for any pre-eminent quality. He seems, indeed, intended to be the patient follower of a ruminating philosopher, or the adulatory and consolatory companion of an old maid; but is now gradually becoming discarded as a pet, and is seldom seen peeping out of a carriage window or basking in a London balcony.

The Comforter, of which a portrait is given at the head of the present chapter, is a rare and beautiful little dog, apparently a cross between the Maltese and King Charles spaniel. His colour is generally white, with black or brown patches; his ears are long, and his head broad on the upper part, with an acute muzzle; the hair is long over the whole body, with the fore legs feathered; his tail is curled, and feathered Page 417 with very long hairs. This is the smallest of any of the distinct races of dogs, and is frequently not above a foot from the tip of the nose to the point of the tail.

"A PUGNACIOUS PAIR."

Page 418

THE TURNSPIT.

How well do I recollect, in the days of my youth, watching the operations of a turnspit at the house of a worthy old Welsh clergyman in Worcestershire, who taught me to read. He was a good man, wore a bushy wig, black worsted stockings, and large plated buckles in his shoes. As he had several boarders, as well as day-scholars, his two turnspits had plenty to do. They were long-bodied, crooked-legged, and ugly dogs, with a suspicious, unhappy look about them, as if they were weary of the task they had to do, and expected every moment to be seized upon to perform it. Cooks in those days, as they are said to be at present, were very cross, and if the poor animal, wearied with having a larger joint than usual to turn, stopped for a moment, the voice of the cook might be heard rating him in no very gentle terms. When we consider that a large solid piece of beef would take at least three hours before it was properly roasted, we may form some idea of the task a dog had to perform in turning a wheel during that time. A pointer has pleasure in finding game, the terrier worries rats with considerable glee, the greyhound pursues hares with eagerness and delight, and the bulldog even attacks bulls with the greatest energy, while the poor turnspit performs his task by compulsion, like a culprit on a treadwheel, subject to scolding or beating if he stops a moment to rest his weary limbs, and is then kicked about the Page 419 kitchen when the task is over. There is a story (it is an old one) of the Bath turnspits, who were in the habit of collecting together in the abbey church of that town during divine service. It is said, but I will not vouch for the truth of the story, that hearing one day the word "spit," which occurred in the lesson for the day, they all ran out of the church in the greatest hurry, evidently associating the word with the task they had to perform.

These dogs are still used in Germany, and her Majesty has two or three of them amongst her collection of these quadrupeds. They are extremely bandy-legged, so as to appear almost incapable of running, with long bodies and rather large heads. They are very strong in the jaws, and are what are called hard-bitten. It is a peculiarity in these dogs that they generally have the iris of one eye black and the other white. Their colour varies, but the usual one is a bluish grey, spotted with black. The tail is generally curled on the back.

As two turnspits were generally kept to do the roasting work of a family, each dog knew his own day, and it was not an easy task to make one work two days running. Even on his regular day a dog would frequently hide himself, so cordially did he hate his prescribed duties. A story is said to have been related to a gentleman by the Duke de Liancourt, of two turnspits employed in his kitchen, who had to take their turns every other day to get into the wheel. One of Page 420 them, in a fit of laziness, hid himself on the day he should have worked, so that his companion was forced to mount the wheel in his stead, who, when his employment was over, began crying and wagging his tail, and making signs for those in attendance to follow him. This was done, and the dog conducted them into a garret, where he dislodged his idle companion, and killed him immediately.

The following circumstance is said to have taken place in the Jesuits' College at La Flèche.

After the cook had prepared his meat for roasting, he looked for the dog whose turn it was to work the spit, but not being able to find him, he attempted to employ for this service another that happened to be in the kitchen. The dog, however, resisted, and, having bitten the cook, ran away. The man, with whom the dog was a particular favourite, was much astonished at his ferocity. The wound he had received was a severe one, and bled profusely, so that it was necessary to dress it. While this was doing, the dog, which had run into the garden, and found out the one whose turn it was to work the spit, came driving him before him into the kitchen, when the latter immediately went of his own accord into the wheel.

Buffon calls the turnspit the *Basset à jambes torses*, but some of the breed are said to have straight legs. Short as they are, the body is extremely strong and heavy in proportion to the height of the dog, and this weight must facilitate the turning of the wheel.

Page 421

THE FOXHOUND.

"Warn'd by the streaming light and merry lark,
Forth rush the jolly clan; with tuneful throats
They carol loud, and in grand chorus joined,
Salute the new-born day.

Then to the copse
Thick with entangled grass, or prickly furze,
With silence lead thy many-coloured hounds
In all their beauty's pride."—Somerville.

It is impossible to enter upon a description of the foxhound without considerable diffidence. Whether we consider the enthusiastic admiration it excites amongst Page 422 sportsmen, the undeviating perseverance and high courage of the animal, its perfect symmetry, and the music of its tongue, which warms the heart and gives life and spirit to man and horse, it must be difficult to do justice to his merits. I will, however, endeavour to do my best; and should I fail, it will not be for want of admiration of the noble animal whose qualifications I am about to illustrate with characteristic anecdotes.

In giving a description of the various breeds of dogs, every one must be aware that by crossing and recrossing them many of those we now see have but little claim to originality. The foxhound, the old Irish wolf-dog, and the colley or shepherd's dog, may, perhaps, be considered as possessing the greatest purity of blood. My opinion respecting the foxhound is partly founded on the following curious fact:—

In Wilkinson's "Manners and Customs of the Egyptians," there is a representation of as varmint a pack of foxhounds as modern eye could wish to see. It is copied from a painting found in the interior of the tomb of the Pharaoh under whom Joseph served. Every individual hound is characteristic of the present breed, with all their courage and animation. Each dog's tail was as an old Irish huntsman, who used to glory in seeing his hounds carry their sterns after the hardest day, once said to his master, "not behind them at all, plaize your honour, but curling out over their shoulders."

Page 423 If the copy be correct, and there is no reason to doubt it, the dog of this breed must be considered of a much more ancient date than is generally supposed. There is every reason to believe that the first dogs came from Asia. Indeed, history, both sacred and profane, confirms this. At all events, the fact just mentioned is sufficiently curious, and may serve to confirm the supposition I have ventured to make of the purity of the blood of our modern foxhound.

A volume might be written on the characteristics of these dogs, both in the kennel and the field, and I will endeavour to illustrate this by a few anecdotes.

It is well known to those who have lived near a kennel, that every morning at the first gleam of light the hounds invariably salute the glorious return of day, by joining simultaneously in a full chorus of voices, 'a musical discord,' called by huntsmen "their morning hymn." This concert does not consist of barking and yapping as many may suppose, but something like the "Hullah system," yet far more sonorous to a sportsman's ear.

Those who have witnessed the process of feeding hounds cannot but acknowledge that it is a most pleasing sight. We see the anxiety depicted in their countenances to detect the huntsman's eye, who calls them singly by name in a low tone of voice, nor does one offer to stir till his time comes. Each dog also takes every day the same position, like children at school, except that all are obedient, and there is no Page 424 noise. His late Majesty George IV., in his younger days, was a constant attendant at the royal kennel at feeding-time, and many of the royal family have also been to see the hounds fed at that place.

Close to the Duke of Beaufort's kennel at Badmington a tame fox was confined, and between it and the foxhounds a great friendship existed. When the hounds were let out they played with the fox, who, on his part, was equally ready to greet them. This reciprocal kindness had continued some time, until one day a hunted fox, much exhausted, ran for shelter into a bush close to the hutch of the tame one. The hounds, in the eagerness of the chase, ran into the latter, mistaking him for the other, and instantly killed him. No sooner, however, were they aware of their having occasioned the death of their old acquaintance, than each hound slunk away, appearing conscious and ashamed of what had been done, nor could they be induced to touch the dead fox when thrown amongst them.

Amongst other curious anecdotes of foxhounds, the following may be mentioned. Some years ago, Sir John Cope had a hound called Clermont, which was in the constant habit, when the pack killed a fox, of taking possession of the animal's head. This he invariably carried in his mouth, as if it was a trophy, and on arriving at the kennel would put it down at the kennel door. In this way he must have imposed a severe task on himself, as the pack had frequently Page 425 twenty miles to go home when the chase was over. The weight was not indeed great; but the dog's mouth being distended the whole time must have made the task anything but a pleasant one.

Some hounds are possessed of extraordinary instinct, which enables them to find their way back to their kennels over country which they had never before traversed. When George III. kept hounds in the Home Park, Windsor, General Manners, one of the equerries, took a hound named Bustler with him in his carriage to London. He remained there a few days, and then travelled to Bloxholm in Lincolnshire, the dog being still his companion inside the carriage. In less than a month, however, Bustler found his way back to Frogmore.

The captain of a vessel informed me that he had once picked up a dog in mid-channel between Brighton and Calais, swimming boldly and strongly towards the French coast. If this dog was endeavouring to make his way back to a beloved master, it was an extraordinary instance of affection.

A few years ago some hounds were embarked at Liverpool for Ireland, and were safely delivered at a kennel far up in that country. One of them, not probably liking his quarters, found his way back to the port at which he had been landed from Liverpool. On arriving at it, some troops were being embarked in a ship bound to that place. This was a fortunate circumstance for the old hound, as during the bustle he Page 426 was not noticed. He safely arrived at Liverpool, and on his old master, or huntsman rather, coming down stairs one morning, he recognised his former acquaintance waiting to greet him.

A similar circumstance happened to some hounds sent by the late Lord Lonsdale to Ireland. Three of them escaped from the kennel in that country, and made their appearance again in Leicestershire.

The love of home, or most probably affection for a particular individual, must be strongly implanted in dogs to induce them to search over unexplored and unknown regions for the being and home they love. Hunger, it might be supposed, would alone stop the ardour of their pursuit, and induce them to seek for nourishment and shelter at a stranger's door. But such is not the case. Hungry, foot-sore, fatigued, and exhausted, the noble and faithful animal presses onward, guided by an instinct which man does not possess, and proving the strength of his love by his indefatigable and ardent exertions. Poor, faithful animal! and is it possible that you are subjected to ill treatment, cruelty, and neglect by those who owe you a large debt of gratitude? Your exertions procure amusement, your watchfulness and fidelity give protection, and neither sickness nor misfortune will induce you to forsake the object of your attachment.

But it is time to resume our anecdotes of foxhounds, and the following is a proof of the high Page 427 courage they so often display, as well as their emulative spirit.

In drawing a strong covert, a young bitch gave tongue very freely, whilst none of the other hounds challenged. The whipper-in rated to no purpose, the huntsman insisted she was wrong, and the whip was applied with great severity, in doing which the lash most unfortunately took the orb of the eye out of the socket. Notwithstanding the excruciating pain she must inevitably have laboured under, the poor suffering animal again flew to the scent, and exultingly proved herself to be right, for a fox having stole away, she broke covert after him unheeded, and continued the chase alone. After much delay and cold hunting the pack at length hit off the chase. At some distance a farmer made a signal with much vehemence to the company, who, upon coming up to him, were informed that they were very far behind the fox, for that a single hound, very bloody about the head, had passed a field from him, and was running breast-high, and that there was little chance of getting up to him. The pack, however, at her coming to a check, did at length get up, and, after some cold hunting, the bitch again hit off the scent, and the fox was killed after a severe run. The eye of the poor but high-spirited dog, which had hung pendent during the chase, was removed by a pair of scissors after the fox was dead.

Page 428 The following is another instance of the persevering strength and spirit of foxhounds:—

A gentleman of the name of Pearson, residing in Essex, had a couple and a half of young and newly-entered hounds. One day they accidentally followed him in his ride, and strayed into a large covert by the roadside, and presently found something which they eagerly hunted. After trying a long time to halloo them off, Mr. Pearson proceeded to Colchester, where his business detained him some hours. Upon his return he heard them in the covert, and found, by some people at work by the side of it, that they had continued running during his absence, and had driven a fox over the field in which they were at work backward and forward several times. Mr. Pearson got as near to them as possible, continuing to give them every encouragement. After hunting the fox a long time in the covert he at last broke, and was killed after a run of some miles. The time these hounds were hunting was seven hours. Hounds have even been known to have continued a chase for ten hours, great part of the time being hard running. A fox was once unkennelled near Boroughbridge in Yorkshire, at twenty-seven minutes past nine, and except half-an-hour taken up in bolting him from a rabbit-burrow, the hounds had a continued run until fourteen minutes past five in the evening, when they killed the fox in good style. During this space of nearly eight hours of most severe run Page 429 ning, several horses died in the field, and others were severely injured.

A hound, the property of Mr. Teasdale of Ousby, Cumberland, during a storm, took the quest of a fox, which he pursued for the extraordinary space of thirty hours, four of which were run within view of some miners, who were employed upon Dalton Fell. The dog and fox were at that time running round the bottom of a hill. The arch dog, still keeping on the side of Reynard which led to his clift in the rock, at last came up to him; but being so much exhausted by his toilsome chase, he was unable to make him his prey for some time, and they lay as if lifeless together. The miners then made up to his assistance; but so ardent was his desire to finish Reynard himself, that he would not suffer them to come near till he had destroyed him.

A foxhound bitch, in the middle of a chase, was taken in labour, and brought forth a puppy. Ardour for the pursuit, united to attachment for her progeny, induced her to snatch it up in her mouth, and follow her companions, with whom she soon came up, and in this interesting situation she continued the whole day,—a discredit to the huntsman, and all who joined in the pursuit, to allow the poor animal to undergo so violent an exercise under such circumstances.

In order to account for the power of endurance which foxhounds are known to possess, it should be Page 430 mentioned that their strength is very great. A well-bred hound has been known to measure as much round the arm of the fore-leg as a moderate-sized horse does below the knee. I was assured of this fact by a well-known huntsman, and it may serve in some measure to account for the following instance of undeviating perseverance in a foxhound, related by Mr. Daniel in his Supplement to his "Rural Sports."

The circumstance took place in the year 1808, in the counties of Inverness and Perth, and perhaps surpasses any length of pursuit known in the annals of hunting. On the 8th of June in that year, a fox and hound were seen near Dunkeld in Perthshire, on the high road, proceeding at a slow trotting pace. The dog was about fifty yards behind the fox, and each was so fatigued as not to gain on the other. A countryman very easily caught the fox, and both it and the dog were taken to a gentleman's house in the neighbourhood, where the fox died. It was afterwards ascertained that the hound belonged to the Duke of Gordon, and that the fox was started on the morning of the 4th of June, on the top of those hills called Monaliadh, which separate Badenoch from Fort Augustus. From this it appeared that the chase lasted four days, and that the distance traversed from the place where the fox was unkennelled to the spot where it was caught, without making any allowances for doubles, crosses, &c., and as the crow flies, exceeded seventy miles.

Page 431 It is a curious fact, that if a foxhound is taken for the first time into a new and strange country, and he is lost, when he returns to his kennel he does so across fields where he had never been before, and not by roads along which he had been taken out. A gentleman who kept foxhounds had an opportunity of observing this. His house and kennel were on the banks of a river, and a new hound accompanied the pack, which went across a bridge near the kennel. He was lost, and came back over the fields direct upon the kennel, and howled when he arrived on the banks of the river. We know but little of the peculiar instinct which thus enables dogs to find their way across a strange country.

Let me here give an anecdote that was communicated to me by the brother of the gentleman to whom it occurred. This gentleman was a rigid Roman Catholic, and his constant companion was a foxhound. As soon as the forty days of Lent began, this dog left his master and came to the house of my informant, some miles distant, where he found food to his liking, and stayed with him during Lent, at the end of which he returned to his owner. He must have measured time very accurately, and has continued the practice for some years.

In the year 1813 some hounds belonging to his late Majesty, George III., were sold to Mr. Walker, of Mitchell Grove, near Worthing. A few weeks after their arrival at that place, one couple of them were sent Page 432 in a stage-waggon to Dr. Willis, then living near Stamford in Lincolnshire. The wagon went through London, and from thence to Dr. Willis's seat. However surprising it may appear, one of these dogs, in less than a month after he had left the kennel near Windsor, found his way back to it. It might be supposed that in this length of time all recollection would have ceased, but such we have seen was not the case.

The circumstance which happened to the late Duke of Northumberland's pack proves the foxhound's eagerness after his game. In 1796 the hounds ran a fox into a very large furze-cover near Alnwick, called Bunker's Hill, where he was lost in an earth which no one knew of. Upon the dogs coming to the kennel two couple and a half of the best of them were missing, and not returning that night, it was thought they had found a fox, and had gone off by themselves in pursuit of him. Several men were sent in search of them to all the earths and crags for twenty miles round, but no tidings could be gained of them. The course where the fox was lost was then searched, and the earth discovered, and in digging about two yards deep, one dog was found; several yards further three more, fast in the ground; and two yards deeper the fifth was dug up. They were all dead.

It is well known to those who served in the Peninsular War, that the late Lord Hill kept a pack of foxhounds while he commanded a division of the army. During a period of repose a fox was unkennelled in the Page 433 neighbourhood of Corja, in Spain. The run was severe for the space of thirty minutes, when the fox, being sharply pressed by the leading hounds, leaped down a precipice of sixty yards perpendicular. Seven couple of the hounds immediately dashed after him, six couple of which were killed on the spot. The remainder of the pack (twenty-two couple) would probably have shared the same fate, had not the most forward riders arrived in time to flog them off, which they did with difficulty, being scarcely able to restrain their impetuosity. The fox was found at the bottom, and covered with the bodies of the hounds.

I might have hesitated to mention the following fact, had it not been witnessed by some well-known sportsmen of the present day.

During a severe chase, and towards the termination of it, when the fox was in view, another fox was seen, to the astonishment of the forward riders, running in the middle of the pack of hounds, perfectly unnoticed by them. It is supposed that the dogs ran over this fox, who, finding himself in the midst of them, probably thought it the safest and wisest plan he could pursue to continue with them till he had an opportunity of making his escape.

In relating anecdotes of foxhounds it is almost unavoidable not to mention fox-hunters, and we know not how we can give to our readers a better notion of the stirring spirit and devotion to their sport, dis Page 434 tinguishing them beyond all other sportsmen, than by offering some extracts from the pen of the late Colonel Cook, a master of hounds, beloved by all who knew him, and venerated by those who hunted with him.

Hounds will not work through difficulties, nor will they exert themselves in that killing sort of manner when they are out of blood. If after all you should, owing to ill-luck and bad weather, be in want of it, the best way is to leave an earth open in a country where you can spare a fox, and where you can without much trouble dig him, give him to the hounds on the earth, and go home. But whatever you do, never turn out a bag-fox; it is injurious to your hounds, and makes them wild and unsteady: besides, nothing is more despicable, or held in greater contempt by real sportsmen, than the practice of hunting bag-foxes. It encourages a set of rascals to steal from other hunts; therefore keep in mind, that if there were no receivers there would be no thieves. What chiefly contributes to make fox-hunting so very far superior to other sports is the wildness of the animal you hunt, and the difficulty in catching him. It is rather extraordinary, but nevertheless a well-known fact, that a pack of hounds, which are in sport and blood, will not eat a bag-fox. I remember hearing an anecdote (when I was in Shropshire many years ago) of the late Lord Stamford's hounds, which I will relate to you as I heard it. Lord Forester, and his brother, Mr. Frank Forester, then boys, Page 435 were at their uncle's for the holidays. A farmer came to inform them a fox had just been seen in a tree. All the nets about the premises were collected, and the fox was caught; but the Squire of Wiley, a sportsman himself, and a strict preserver of foxes, sent the fox immediately to Lord Stamford by one of his tenants, that he might be informed of the real circumstance. The next day the hounds were out, and also the Squire's tenant; they had drawn some time without finding, when the farmer reminded his Lordship of the fox caught. 'Do you think,' said he, 'I will allow my hounds to hunt a bag-fox? I should never be forgiven by my huntsman!' At last, after drawing several coverts without finding, his Lordship gave his consent (but it was to be kept a great secret), and the bag was to be touched upon the ground in a line for a covert they were going to draw, to have the appearance of a disturbed fox, and the fox to be turned down in it.

On going to covert, a favourite hound, called Partner, feathered on the scent. The huntsman exclaimed in ecstacy, 'Old Partner touches on him; we shall certainly find in the next covert.' They found the bag-fox, and had a tolerable run; but when they killed him, not a hound would eat him! 'Now, Sir,' said his lordship to the farmer, 'you have deceived the huntsman and the field, but you cannot deceive my hounds.'

Page 436 Next to turning out bag-men, lifting of hounds is the most prejudicial. They should seldom be taken 'off their noses,' nothing is gained by it in the end; hounds that are seldom lifted will kill more foxes in the course of a season than those that frequently are. Some years ago, when hunting with the Duke of Grafton's hounds in Suffolk, they came to a check all in a moment, at a barn near some cross-roads; they were left alone, and made a fling of themselves, in a perfect circle, without hitting the scent; many gentlemen exclaimed, 'It is all over now, Tom; the only chance you have is to make *a wide cast*.' 'No,' answered the huntsman, 'if the fox is not in that barn, my hounds ought to be hung.'

Dick Foster, the whipper-in, now huntsman to Mr. Villebois (and a very good one he is), was ordered to dismount and see if he could discover the fox; he returned and said he was *not* there.' Tom Rose still was positive; at last he was viewed on a beam in the barn, and they killed him, after a further run of about a mile. I mention this trivial circumstance to show you clearly, that if the hounds had been hurried up either of the roads on a wide cast, made by an ignorant huntsman, the fox would inevitably have been lost.

Were I to have some sporting friends coming to see my hounds in the field, I should prefer going away *close at him* for twenty minutes, then a short Page 437 check, to bring the hounds to a hunting scent, and a quick thing at last, and run into him, in order that my friends might be convinced the hounds could *hunt* as well as run; for of this I am certain, if they cannot do *both*, they merit not the name of fox-hounds.

HEAD OF A FAVORITE FOX-HOUND.

HOUNDS IN A BATH.

Page 438

THE BEAGLE.

The beagle may be mentioned as a sort of foxhound in miniature, and nothing can well be more perfect than the shape of these small dogs. But how different are they in their style of hunting! The beagle, which has always his nose to the ground, will puzzle for a length of time on one spot, sooner than he will leave the scent. The foxhound, on the contrary, full of life, spirit, and high courage, is always dashing and trying forward. The beagle, however, has extraordinary perseverance, as well as nicety of scent, and also a liveliness of manner in hunting, which, joined to its musical and melodious note, will always afford pleasure to the lovers of the chase, or at least to those who are unable to undertake the more exciting sport of fox-hunting. In rabbit-shooting, in gorse and thick cover, Page 439 nothing can be more cheerful than the beagle; and they have been called rabbit-beagles from this employment, for which they are peculiarly qualified, especially those dogs which are somewhat wire-haired.

In the reign of Queen Elizabeth a race of beagles had been bred so small, that a pack of them could be carried out to the field in a pair of panniers. That Princess is said to have had little *singing beagles,* a single one of which could be placed in a man's glove, and they probably at this time received the name of *lap-dog* beagles. Dryden, in his "Fables," alludes to these dogs as follows:—

"The graceful goddess was array'd in green;
About her feet were little beagles seen,
That watch'd with upward eyes the motions of their queen."

Pope also mentions them,—

"To plains with well-bred beagles we repair,
And trace the mazes of the circling hare."

Page 440

THE MASTIFF.

"Great Brittain was so noted for its Mastiffs, that the Roman Emperors appointed an Officer in this Island, with the title of Procurator Cynegii, whose sole business was to breed, and transmit from hence to the Amphitheatre, such as would prove equal to the combats of the place:

Magnaque taurorum fracturi colla Britanni."

This noble dog, which, like the bull-dog, is supposed to be an original breed peculiar to this country, is now seldom to be met with in its pure state, it having been crossed and recrossed with other dogs. Perhaps Page 441 the finest specimen now to be found is one at Chatsworth (where also is to be seen a noble Alpine mastiff). It is a dog of gigantic size, of a yellowish colour, with a black muzzle. There is also another at Elvaston Castle in Derbyshire, not so large as the one at Chatsworth, but apparently of the true breed, and for which we believe Lord Harrington gave the sum of fifty guineas.

These dogs are brave, faithful to their trust in an extraordinary degree, and have a noble disposition.

Their strength also is very great, and their bark deep and loud. Sir Walter Scott's remarks on the character of the dog may be well applied to the mastiff,—"The Almighty, who gave the dog to be the companion of our pleasures and our toils, hath invested him with a nature noble and incapable of deceit. He forgets neither friend nor foe—remembers, and with accuracy, both benefit and injury. He hath a share of man's intelligence, but no share of man's falsehood. You may bribe a soldier to slay a man with his sword, or a witness to take life by false accusation, but you cannot make a dog tear his benefactor. He is the friend of man, save when man justly incurs his enmity."

The mastiff, indeed, usually shows a remarkable and peculiar warmth in his attachments; and, on the other hand, he will evince his dislike in the strongest manner. It has been observed of him, that if he is once severely corrected or insulted, it is almost im Page 442 possible to eradicate the feeling from his memory, and it is no less difficult to attain a reconciliation with him. He seems conscious of his own strength, power, and authority, and will seldom condescend to lower his dignity by servile fawning; while he appears to consider his services as only befitting a trust of the highest importance. He is naturally possessed of strong instinctive sensibility, speedily obtains a knowledge of all the duties required of him, and discharges them with the most punctual assiduity. His vigilance is very striking. He makes regular rounds of the premises committed to his care, examines every part of them, and sees that everything is in a state of perfect security. During the night he will give a signal of his presence by repeated barkings, which are increased upon the least cause of alarm. Unlike the bull-dog, the mastiff always warns before he attacks. His voice is deep and powerful in tone.

Such is the animal of which I now propose to give a few characteristic anecdotes.

About the year 1742, a lady, who resided in a lone house in Cheshire, permitted all her servants, except one female, to go to a supper and dance at a Christmas merry-meeting, held at an inn about three miles distant, and kept by the uncle of the maid who had remained in the house with her mistress. The servants were not expected back till the morning; consequently the doors and windows were, as usual, secured, and the lady and her servant were going to bed, when they Page 443 were alarmed by the voice of some persons apparently attempting to break into the house. Fortunately a great mastiff dog, named Cæsar, was in the kitchen, and set up a tremendous barking, which, however, had not the effect of intimidating the robbers. The maid-servant distinctly heard that the attempt to enter the house was made by the villains endeavouring to force a way through a hole under the sunk story in the adjoining back-kitchen or scullery. Being a young woman of courage, she went towards the spot, accompanied by the dog, and patting him on the back, exclaimed, "At him, Cæsar!" The dog made a furious attack on the person who seemed to be at the hole, and gave something a violent shake, when all became quiet, and the animal returned to her with his mouth all besmeared with blood. She afterwards heard some little bustle outside of the house, which soon was stilled. The lady and servant sat up until morning, without farther molestation, when, on going into the court, a quantity of blood was found on the outside of the wall. The other servants, on their return, brought word to the maid that her uncle, the innkeeper, had died suddenly during the course of the night—they understood of a fit of apoplexy—and was intended to be buried that day. The maid got leave to go to the funeral, and was surprised to find the coffin on her arrival screwed down. She insisted on taking a last view of the body, which was most unwillingly granted; when, to her great surprise and horror, she found his Page 444 death had been occasioned from his throat being torn open. What had happened the evening before immediately rushed to her imagination, and it appeared too evident to her that she had been the innocent cause of her uncle's death; and, upon further inquiry, it was proved that he and one of his servants had formed the design of robbing the house and murdering the lady, in her unprotected condition, during the absence of her servants; but, by the watchfulness and courage of her dog, their design was frustrated.

An anecdote is related of a mastiff, who, in the reign of Queen Elizabeth, when Lord Buckhurst was ambassador at the Court of Charles the Ninth, alone and unassisted, successively engaged a bear, a leopard, and a lion, and pulled them all down.

Very extraordinary stories have been told of these and some other kinds of dogs discovering and circumventing plans to injure the persons of their masters, in which it is difficult to place implicit credit. We give one of the most marvellous of these anecdotes, as it is usually related:—

Sir H. Lee, of Ditchley, in Oxfordshire, ancestor of the late Earls of Lichfield, had a mastiff which guarded the house and yard, but had never met with any particular attention from his master. In short, he was not a favourite dog, and was retained for his utility only, and not from any partial regard.

One night, as Sir Harry was retiring to his chamber, attended by his favourite valet, an Italian, the Page 445 mastiff silently followed them up-stairs, which he had never been known to do before, and, to his master's astonishment, presented himself in the bed-room. Being deemed an intruder, he was instantly ordered to be turned out; which, being complied with, the poor animal began scratching violently at the door, and howling loudly for admission. The servant was sent to drive him away. Discouragement, however, could not check his intended labour of love; he returned again, and was more importunate to be let in than before. Sir Harry, weary of opposition, though surprised beyond measure at the dog's apparent fondness for the society of a master who had never shown him the least kindness, and wishing to retire to rest, bade the servant open the door, that they might see what he wanted to do. This done, the mastiff, with a wag of the tail, and a look of affection at his lord, deliberately walked up, and crawling under the bed, laid himself down, as if desirous to take up his night's lodging there.

To save farther trouble, and not from any partiality for his company, this indulgence was allowed. The valet withdrew, and all was still. About the solemn hour of midnight the chamber door opened, and a person was heard stepping across the room. Sir Harry started from sleep; the dog sprung from his covert, and seizing the unwelcome disturber, fixed him to the spot. All was dark: Sir Harry rang his bell in great trepidation, in order to procure a light. The person Page 446 who was pinned to the floor by the courageous mastiff roared for assistance. It was found to be the favourite valet, who little expected such a reception. He endeavoured to apologise for his intrusion, and to make the reasons which induced him to take this step appear plausible; but the importunity of the dog, the time, the place, the manner of the valet, raised suspicions in Sir Harry's mind, and he determined to refer the investigation of the business to a magistrate.

The perfidious Italian, alternately terrified by the dread of punishment and soothed by the hope of pardon, at length confessed that it was his intention to murder his master, and then rob the house. This diabolical design was frustrated solely by the unaccountable sagacity of the dog and his devoted attachment to his master. A full-length picture of Sir Harry, with the mastiff by his side, and the words, "More faithful than favoured," is still preserved among the family pictures.

Presentiments of approaching danger, such as those now related, are to be traced only to the animal's close observation and watchful jealousy of disposition. Looks, signs, and movements are noticed by him which escape an ordinary observer. The idea that dogs have presentiments of death, and howl on such occasions, is a superstition now all but vanished.

In October 1800, a young man going into a place of public entertainment at Paris, was told that his dog (a fine mastiff) could not be permitted to enter, Page 447 and he was accordingly left with the guard at the door. The young man was scarcely entered into the lobby, when his watch was stolen. He returned to the guard, and prayed that his dog might be admitted, as, through his means, he might discover the thief: the dog was suffered to accompany his master, who intimated to the animal that he had lost something; the dog set out immediately in quest of the strayed article, and fastened on the thief, whose guilt on searching him was made apparent: the fellow had no less than six watches in his pocket, which being laid before the dog, he distinguished his master's, took it up by the string, and bore it to him in safety.

At the castle of a nobleman in Bohemia, a large English mastiff was kept, that never failed to go every Sunday to the village church. The other dogs in the neighbourhood used to follow him thither, so that the church was often full of these animals. This being considered a nuisance, orders were given by the magistrates, at one of the petty courts held for regulating the affairs of the village, that the inhabitants should be enjoined to keep all their dogs locked up every Sunday during the time of divine service. The magistrate who presided in this court said, in a loud and authoritative tone of voice, "I will suffer no dogs in the church; let me not see one there in future." The mastiff happened to be lying under the table in the court when these words were spoken, to which he appeared to listen with great attention. On the ensuing Page 448 Sunday the dog rose at an early hour, ran from house to house through the village, barking at the windows, and at last took his station before the church-door, to see whether any of his companions would venture to approach it, notwithstanding the prohibition. Unfortunately one of them appeared. The mastiff immediately fell upon him with the utmost fury, bit him to death, and dragged him out into the street. He continued in the same manner for several subsequent Sundays to stand sentinel, without ever entering the church.

Captain Brown gives an interesting instance of the gentleness of a mastiff towards a child. He says that a large and fierce mastiff, which had broken his chain, ran along a road near Bath, to the great terror and consternation of those whom he passed. When suddenly running by a most interesting boy, the child struck him with a stick, upon which the dog turned furiously on his infant assailant. The little fellow, so far from being intimidated, ran up to him, and flung his arms round the neck of the enraged animal, which instantly became appeased, and in return caressed the child. It is a fact well known, that few dogs will bite a child, or even a young puppy. Captain Brown adds, that he possesses a mastiff, which will not allow any one of his family to take a bone from him except his y-oungest child.

A chimney-sweeper had ordered his dog, a mastiff crossed with a bull-dog, to lie down on his soot-bag, Page 449 which he had placed inadvertently almost in the middle of a narrow back-street in the town of Southampton. A loaded coal-cart passing by, the driver desired the dog to move out of the way. On refusing to do so, he was scolded, then beaten, first gently, and afterwards with a smart application of the cart-whip, but all to no purpose. The fellow, with an oath, threatened to drive over the dog, and he did so, the faithful animal endeavouring to arrest the progress of the wheel by biting it. He thus allowed himself to be killed sooner than abandon his trust.

A mastiff-dog, who owed more to the bounty of a neighbour than to his master, was once locked by mistake in the well-stored pantry of his benefactor for a whole day, where milk, butter, bread, and meat, within his reach, were in abundance. On the return of the servant to the pantry, seeing the dog come out, and knowing the time he had been confined, she trembled for the devastation which her negligence must have occasioned; but, on close examination, it was found that the honest creature had not tasted of anything, alt-hough, on coming out, he fell on a bone that was given to him, with all the voraciousness of hunger.

These dogs are alive to injuries, and not slow in resenting them.

A carrier had a mastiff remarkable for his sagacity. It happened unfortunately one day, that one of the waggon-horses trod acciden-tally upon him in the yard. The dog became furious, and would have attacked the Page 450 horse had he not been prevented. It was usual for the dog to remain with the horses at night in the stable. After the men had retired, the mastiff selected out the animal which had trod upon him, and, no doubt, would have put an end to his existence, had not the carters, who were at hand, hearing an unusu-al noise, come to his assistance.

The widow of a farmer had two mastiffs, which, from their fier-ceness, rendered some precaution necessary in approaching the house. Their mistress was taken suddenly ill and died, and in the afternoon of her death the benevolent wife of the clergyman of the parish called to see if she could render any assistance. After knocking in vain at the front door, she went to the back of the house with fear and trembling. On entering the kitchen, to her dismay she saw the two dogs on the hearth. They appeared, however, to be sensible of what had taken place, for they only lifted up their heads mournfully, looked at the intruder, and resumed their former attitude.

My neighbour, Mr. Penrhyn, has two noble mastiffs of the Lyme breed, which I believe is now nearly extinct. It is probably, however, preserved by Thomas Leigh, Esq. of Lyme Park, in Cheshire, who has also the wild breed of cattle, now only, I believe, found at Lyme Park, and at Chillington, in Yorkshire, the seat of Lord Tankerville. There is a story current at Lyme Park, that some years ago a dog of the breed in question, whilst walking with the steward in the park, took Page 451 offence at one of the wild bulls, and would instantly have attacked it, but was with difficulty restrained by the steward. The dog returned home, evidently bearing the offence in mind, and the next morning, the steward, seeing him covered with blood, suspected something amiss, and on going into the park, found that not only the bull, but two cows had been worried by him.

A mastiff belonging to a tanner had taken a great dislike to a man, whose business frequently brought him to the house. Being much annoyed at his antipathy and fearful of the consequences, he requested the owner of the dog to endeavour to remove the dislike of the animal to him. This he promised to do, and brought it about in the following manner, by acting on the noble disposition of the dog. Watching his opportunity, he one day, as if by accident, pushed the dog into a well in the yard, in which he allowed it to struggle a considerable time. When the dog seemed to be getting tired, the tanner desired his companion to pull it out, which he did. The animal, on being extricated, after shaking himself, fawned upon his deliverer, as if sensible that he had saved his life, and never molested him again. On the contrary he received him with kindness whenever they met, and often accompanied him a mile or two on his way home.

A personal friend of the writer's, some time since, on a visit at a gentleman's house in the country, was taking a moonlight walk through the shrubbery and pleasure-grounds, when he was startled by a noise Page 452 behind him; on turning his head, he perceived a large mastiff, which was ordinarily let loose as evening closed, and which had tracked him through the grounds. The dog with a fierce growl roughly seized him; our friend wisely deemed passive obedience and non-resistance the most prudent if not the most courageous part for him to play, and was unceremoniously led back through the grounds to the hall-door; here he was relieved by the master of the house. Subsequently assured that he had no cause to fear, he repeated his walk; the dog was again at his side, but walked quietly with him, and acknowledged in the usual way his words of conciliation. On these instances of sagacity (sagacity of a kind very different from that displayed by the shepherd's dog or the setter) there needs no comment.

A gentleman in Ireland had a mastiff which was kept to guard his premises. A small dog, belonging to a poor man who came to the house on business, had barked at and annoyed him, but he was obliged to submit to the insult at the time with sullen patience, as his chain prevented him from taking any immediate revenge. A few evenings afterwards, however, he contrived to escape from the back-yard, and immediately made his way to the cabin of the cur's master. Finding the door open, *more Hibernicorum*, he entered without even a premonitory growl, to the dismay of the humble inmates, who were eating their supper of potatoes and milk, seized the offender, and killed it.

Another mastiff behaved in a very different manner. Page 453 He had also been annoyed by a little cur as he passed along the streets, which he bore with great patience for a long time; at last his persecutor became so troublesome that he could bear it no longer. He, therefore, one day caught his contemptible adversary by the neck, carried him to the edge of a wharf, and dropped him gently into the water. [S]

The instinctive appreciation of the nature of property as shown in dogs is exemplified in the following instance:—A lady at Bath, walking out one day, was impeded in her progress by a strange mastiff dog. She became alarmed, and at the same time perceived that she had lost her veil. Upon retracing her steps, the dog went on before her, till the lost article was discovered; and as soon as it was picked up, the animal hastened after his own master.

Page 454

THE BULL-DOG.

"The heroes of a bull-fight, and the champions of a cock-fight, can produce but few, if any, disciples brought up under their tuition, who have done service to their country, but abundant are the testimonies which have been registered at the gallows of her devoted victims, trained up to the pursuits of bull-baiting."—Dr. Barry.

The bull-dog has been called the most courageous animal in the world. He is low in stature, although remarkably deep-chested, strong, and muscular. From the projection of his under jaw, which occasions his teeth always to be seen, and from his eyes being distant Page 455 from each other, and somewhat prominent, he has an appearance which would prevent a stranger from attempting any familiarity with him. He is, however, a dog capable of strong attachment to his master, whom he is at all times ready to defend. His strength is so great, that in pinning a bull, one of this breed of dogs has been known, by giving a strong muscular twist of his body, to bring the bull flat on his side. In consequence also of his strength, high courage, and perseverance, a bull-dog has gone a greater distance in swimming than any other dog has been known to do.

It is universally known amongst the lovers of bull-dogs, that when once exasperated by an opponent or encouraged by the owner, no pain or punishment will induce him to swerve from his purpose, or in the least relax the violence of his endeavours to subdue whatever may be the object of his dislike or resentment. Amidst the many instances which might be adduced in support of this assertion, we shall notice one which is well-authenticated. Some years since, when bull-baiting was more common than in the present improved state of civilization, a juvenile amateur, at an entertainment of this kind in the north of England, confident in the courage and purity of blood in his bull-dog, laid a wager "that he would at four distinct intervals deprive the animal of one of his feet by amputation, and that after every individual deprivation he should still attack the bull with his previous ferocity; and that, lastly, he should continue to do so Page 456 upon his stumps." Shocking as the recital must prove to the feelings of every reader, the experiment was made, and the dog continued to seize the bull with the same eagerness as before. In a match which was made for the purpose, one of these animals fought and beat two powerful Newfoundland dogs.

It must be a matter of congratulation to every humane person, that the barbarous and cruel custom of bull-baiting no longer exists in this country. That it tended to brutalize the working classes, whatever its advocates may have stated to the contrary, cannot be doubted. In the part of Staffordshire in which I formerly resided, and where the custom was extremely prevalent, idleness, drunkenness and profligacy, were conspicuous amongst those who kept bull-dogs. Even females might be seen at a bull-baiting, in their working dresses as they came out of a factory, their arms crossed and covered with their aprons, standing to enjoy the sport, if such it could be called.

The breed of dogs kept by the persons referred to was said to be of the purest kind, and large sums were frequently given for them. Lord Camelford purchased one for eighty guineas; forty and fifty pounds was no uncommon price for one. These dogs would appear to have a natural antipathy to the bull, as puppies will attack them when only a few months old, and if permitted to continue the combat, will suffer themselves to be destroyed rather than relinquish the contest. A well-bred dog always attacks the bull in front, and Page 457 endeavours to seize on the lip as the most sensitive part.

A nobleman had a favourite bull-dog, which was his constant companion in his carriage to and from his seat in Scotland for many years. The dog was strongly attached to his master, and was gentle and inoffensive. As he grew old, it was determined to leave him in London. The carriage came to the door, his master entered it, and drove off, taking another dog for his companion. The packing—the preparations—had all been witnessed by the faithful bull-dog, who was evidently aware that he had been deserted by the only being he loved. From that moment he became melancholy. He refused to eat, and notwithstanding all the care taken of him, he pined and died.

A bull-dog, not many years since, saved a shipwrecked crew by towing a rope from the vessel to the shore, after two fine Newfoundland dogs had perished in the attempt. This success may be attributed to his indomitable courage, which prevented him from giving up his exertions while life remained.

I remember many years ago hearing of some robberies, which took place by means of a bull-dog in the neighbourhood of London, one of which was near my own residence. A gentleman in riding home one winter's evening, had one of the hocks of his horse seized, as he was trotting along the road, by a bull-dog, who kept his hold, and brought the horse to the ground. A man then came up, and robbed the gentleman of his purse.

Page 458 It was common in Staffordshire, before young dogs were able to cope with a bull, to practise them with a man, who stood proxy for the bull. On one occasion of this sort, Mr. *Deputy* Bull being properly staked, began to perform his part by snorting and roaring lustily. The dog ran at him, but was repulsed,—the courage of the animal, however, increased with every struggle, and at last he seized his biped antagonist by the cheek, who, with rueful countenance, endured it for some time, till at length he was compelled to cry out to his companion to take the dog off; but he, unwilling to damp the courage of his *élève*, vociferated, "*Woot* spoil the pup, *mun*?—let 'em taste *bloode* first!"

Bull-dogs are now much less common than they were. A cross breed between them and a good terrier is said to produce better fighters and harder biters than the pure bull-dog. If one of these dogs is crossed with a greyhound, the offspring is found to be too courageous, and from this cause in attacking deer they have been frequently killed.

Page 459

THE DALMATIAN OR COACH-DOG.

This dog, says Mr. Bewick, has been erroneously called the Danish dog by some authors, and by Buffon the harrier of Bengal; but his native country is Dalmatia, a mountainous district on the

Adriatic coast. He has been domesticated in Italy for upwards of two centuries, and is the common harrier of that country.

The Dalmatian is also used there as a pointer, to which his natural propensity more inclines him than to be a dog of the chase: he is said to be easily broken, and to be very staunch. He is handsome in shape, something between the British foxhound and English pointer; his head more acute than that of the latter, and something longer: his general colour white, and his whole body and legs covered with small irregular-sized black or reddish-brown spots. The pure breed has tanned cheeks and black ears. He is much smaller than the large Danish dog. A singular opinion prevailed at one time in this country, that this beautiful dog was rendered more handsome by having his ears cropped: this barbarous fancy is now fast dying away.

The only use to which this elegant dog is applied is as an attendant upon a carriage, for which the symmetry of his form and beauty of his skin peculiarly fit him. He familiarises readily with horses, and is therefore invariably entrusted to the stables. A most erroneous notion has long prevailed that neither Page 460 this nor the great Danish dog has the sense of smell. They have been indiscriminately called the Coach-dog.

Mr. Dibdin, in his "Tour through England," says, "I took with me last summer one of those spotted dogs called Danish, but the breed is Dalmatian. It was impossible for anything to be more sportive, yet more inoffensive, than this dog. Throughout the mountainous parts of Cumberland and Scotland his delight was to chase the sheep, which he would follow with great alertness even to the summits of the most rugged steeps; and when he had frightened them, and made them scamper to his satisfaction (for he never attempted to injure them), he constantly came back wagging his tail, and appearing very happy at those caresses which we, perhaps absurdly, bestowed upon him.

"About seven miles on this side of Kinross, in the way from Stirling, he had been amusing himself playing these pranks, the sheep flying from him in all directions, when a black lamb turned upon him, and looked him full in the face; he seemed astonished for an instant, but before he could rally his resolution, the lamb began to paw and play with him. It is impossible to describe the effect this had upon him; his tail was between his legs, he appeared in the utmost dread, and slunk away confused and distressed: presently his new acquaintance invited him, by all manner of gambols, to be friendly with him. What a moment for Pythagoras or Lavater! Gradually overcoming his fears, he accepted this brotherly challenge, and they Page 461 raced away together, and rolled over one another like two kittens. Presently appeared another object of distress. The shepherd's boy came to reclaim his lamb; but it paid no attention except to the dog, and they were presently at a considerable distance. We slackened our pace for the convenience of the boy, but nothing would do; we could no more call off the dog than he could catch the lamb. They continued sporting in this manner for more than a mile and a half. At length, having taken a circuit, they were in our rear; and after we had crossed a small bridge, the boy with his pole kept the lamb at bay, and at length caught him; and having tied his plaid round him, it was impossible for him to escape. Out of fear of the boy, and in obedience to us, the dog followed reluctantly; but the situation of the lamb all this time cannot be pictured; he made every possible attempt to escape from the boy, even at the risk of tumbling into the river, rather than not follow the dog. This continued till the prospect closed, and we had lost sight of our new ally, whose unexpected offer of amity to the Dalmatian seemed ever after to operate as a friendly admonition, for from that day he was cured of following sheep."

Lord Maynard, some years since, lost a coach-dog in France, which he in vain endeavoured to find. He returned to England, where he had not long arrived before the dog appeared; but the mode of his return remained for ever unexplained, though it is more than probable that the dog's sagacity, when he had made his Page 462 escape from confinement, prompted him to go to the sea-coast, where he found means to get on board some vessel bound for the opposite shore.

The late Mr. Thomas Walker, of Manchester, had a small Dalmatian dog, which was accustomed to be in the stable with two of his carriage-horses, and to lie in a stall with one of them, to which he was particularly attached. The servant who took care of the horses was ordered to go to Stockport (which is distant about seven miles), upon one of the horses, and took the one above mentioned (the favourite of the dog), with him, and left the other with the dog in the stable; being apprehensive lest the dog, which was much valued by his master, should be lost upon the road. After the man and horse had been gone about an hour, some person coming accidentally into the stable, the dog took the opportunity of quitting his confinement, and immediately set off in quest of his companion. The man, who had finished the business he was sent upon, was just leaving Stockport, when he was surprised to meet the dog he had left in the stable, coming with great speed down the hill into the town, and seemed greatly rejoiced to meet with his friendly companion, whom he had followed so far by scent. The friendship between these animals was reciprocal; for the servant, going one day to water the carriage-horses at a large stone trough, which was then at one end of the exchange, the dog as usual accompanying them, was attacked by a large mastiff, and in danger of being Page 463 much worried, when the horse (his friend), which was led by the servant with a halter, suddenly broke loose from him, and went to the place where the dogs were fighting, and with a kick of one of his heels struck the mastiff from the other dog clean into a cooper's cellar opposite; and having thus rescued his companion, returned quietly with him to drink at the conduit.

THE GREAT DANISH DOG.

Buffon was of opinion that this variety, which is chiefly found in Denmark, Russia, and Northern Germany, is only the Mâtin (the usual sheep-dog of France) transported into a northern latitude. The colour of this dog is generally white, marked all over his body with black spots and patches, in general larger than those of the Dalmatian, of which some have supposed him to be a congener. His ears are for the most part white, while those of the Dalmatian are usually black.

The great Danish dog is a fine sprightly animal, but is of little use either for sporting or watching. Like the Dalmatian, he is chiefly used in this country as an attendant on carriages, to which he forms an elegant appendage.

Mr. Johnson, a traveller from Manchester, on his route through Scotland on horseback, was benighted, and coming to a small public-house on the road, he Page 464 thought it better to take up his lodgings there, if possible, than to proceed further that night. On entering the house, he found only an old woman, who, to his inquiries, answered she would accommodate him with a bed, and provide for the horse in a small shed, if he would assist her in carrying hay and litter, as there was no other person then in the house. This was readily agreed to by Mr. Johnson, who, after having done so, and taken a little refreshment, was shown by the old woman to his bedroom.

A large Danish dog, which accompanied him on his journey, offered to go up to the room with him, which the old woman strongly objected to, but Mr. Johnson firmly persisted in having him admitted. The dog, on entering the room, began to growl, and was altogether very unruly. His master in vain attempted to quiet him,—he kept growling and looking angrily under the bed, which induced Mr. Johnson to look there likewise, when, to his utter astonishment, he saw a man concealed at the farther end. On encouraging the dog, he sprang immediately at him, whilst Mr. Johnson seized his pistols, and presenting one at the stranger, who had a large knife in his hand, and was struggling with the dog, declared he would instantly shoot him if he made further resistance. The man then submitted to be bound, and acknowledged that his intention was to rob and murder Mr. Johnson, which was thus providentially prevented by the wonderful sagacity of his faithful dog. Mr. Page 465 Johnson, after securely binding the man and fastening the door, went (accompanied by his dog) to the shed where his horse was left, which he instantly mounted, and escaped without injury to the next town, where he gave to a magistrate a full account of the murderous attempt, and the culprit was taken into custody and afterwards executed.

A gamekeeper belonging to the castle of Holstein (in Denmark), returned one evening from a long and fatiguing chase, and deposited the game in the larder, without being aware that he had locked up his dog at the same time. Business of importance unexpectedly called him away immediately afterwards, and he did not return for five days; when, mindful of his game, he went to the larder, and beheld his dog stretched dead at the door. The gamekeeper stood extremely affected; but what were his sensations, when he saw on the table eleven brace of partridges, and five grouse untouched! This admiration increased his grief, when he found the poor dog had suffered starvation rather than transgress his duty.

At a convent in France, twenty paupers were served with a dinner at a certain hour every day. A mâtin dog belonging to the convent did not fail to be regularly present at this repast, to receive the scraps which were now and then thrown to him. The guests, however, were poor and hungry, and of course not very wasteful, so that their pensioner did little more than scent the feast, of which he would fain have partaken. The Page 466 portions were served by a person at the ringing of a bell, and delivered out by means of what in religious houses is termed a *tour*—a machine like the section of a cask, that, by turning round on a pivot, exhibits whatever is placed on the concave side, without discovering the person who moves it. One day this dog, who had only received a few scraps, waited till the paupers were all gone, took the rope in his mouth, and rang the bell. His stratagem succeeded. He repeated it the next day with the same good fortune. At length the cook, finding that twenty-one portions were given out instead of twenty, was determined to discover the culprit. In doing which he had no great difficulty; for, lying in wait, and noticing the paupers as they came for their different portions, and that there was no intruder except the dog, he began to suspect the truth; which he was confirmed in when he saw the animal continue with great deliberation till the visitors were all gone, and then pull the bell. The matter was related to the community; and to reward him for his ingenuity, the dog was permitted to ring the bell every day for his dinner, on which a mess of broken victuals was always afterwards served out to him.

THE CUR DOG.

Almost every dog which is cross-bred is ranked as a cur dog or mongrel, but that which is specially described by Youatt, is the shepherd's dog crossed Page 467 with the terrier, and is nearly smooth; but he is considerably longer in the legs in proportion to the size of his body, is stronger in the make, has half-pricked ears, is generally black and white, although sometimes all black, and has rather a short tail. In the north of England and southern counties of Scotland great attention is paid to the breeding of this dog, and to breaking him in for driving and tending cattle, which he does with great intelligence; indeed his sagacity in everything is uncommonly great, and he is very trusty. These dogs bite very keenly, and always make their attack at the heels of cattle, who, on this account, having no defence against them, are quickly compelled to run.

The cur has long and somewhat deservedly obtained a very bad name as a bully and a coward; and certainly his habit of barking at everything that passes, and flying at the heels of the horse, renders him often a very dangerous nuisance. He is, however, valuable to the cottager; he is a faithful defender of his humble dwelling; no bribe can seduce him from his duty; and he is a useful and an effectual guard over the clothes and scanty provisions of the labourer, who may be working in some distant part of the field. All day long he will lie upon his master's clothes seemingly asleep, but giving immediate warning of the approach of a supposed marauder. He has a propensity, when at home, to fly at every horse and every strange dog; and of young game of every kind there is not a more ruthless destroyer than the village cur.

Page 468 The following story is strictly authentic: — "Not long ago a young man, an acquaintance of Lord Fife's coachman, was walking, as he had often done, in his lordship's stables at Banff. Taking an opportunity when the servants were not regarding him, he put a bridle into his pocket. A Highland cur that was generally about the stables observed the theft, and immediately began to bark at him; and when he got to the stable door would not let him pass, but held him fiercely by the leg to prevent him. As the servants had never seen the dog act thus before, and the same young man had been often with them, they could not imagine what could be the reason of the dog's conduct. However, when they perceived the end of a valuable bridle peeping out of the young man's pocket they were able to account for it, and on his giving it up the dog let go his hold and allowed him to pass."

"I recollect," says Mr. Hall, "when I passed some time at the Viscount Arbuthnot's at Hatton, in the parish of Marykirk, one of his lordship's estates, that when the field-servants went out one morning they found a man whom they knew, and who lived a few miles' distance, lying on the road a short way from the stable with a number of bridles, girths, &c. &c. near him, and the house-dog, which was of the Highland breed, lying also at his ease, holding the seat of the man's breeches in his mouth. The man confessed his crime, and told them that the log had struggled with Page 469 him, and held him in that position for five hours; but that immediately after the servants came up he let go his hold."

The following anecdote is well known. In London, a few years since, a box, properly directed, was sent to a merchant's shop to lie there all night, and be shipped off with other goods next morning. A dog, which accidentally came into the shop with a customer, by smelling the box, and repeatedly barking in a peculiar way, led to the discovery that it did not contain goods, but a fellow who intended to admit his companions and plunder the shop in the night-time.

John Lang, Esq., deputy-sheriff of Selkirk, had a female cur big with pups, which on one occasion, when out in the fields attending the cattle, was taken in travail, and pupped on the moor. She concealed her litter in a whin-bush, brought the cattle home at the usual time with the utmost care, and, having delivered her charge, returned to the moor and brought home the puppies one by one. Mr. Lang, with that humanity which marks his character, preserved the whole litter, that he might not give the least cause of pain to so faithful and so affectionate an animal.

In Lambeth Church there is a painting of a man with a dog on one of the windows. In reference to this, we learn by tradition that a piece of ground near Westminster Bridge, containing one acre and nineteen roods (named Pedlar's Acre), was left to this parish by a pedlar, upon condition that his picture, and that of Page 470 the dog, should be perpetually preserved on painted glass on one of the windows of the church, which the parishioners have carefully performed. The time of this gift was in 1504, when the ground was let at 2*s.* 8*d.* per annum; but in the year 1762 it was let on lease at 100*l.* per year, and a fine of 800*l.*; and is now worth more than 250*l.* yearly. The reason alleged for the pedlar's request is, that being very poor, and passing the aforementioned piece of ground, he could by no means get his dog away, which kept scratching a particular spot of earth, until he attracted his master's notice; who going back to examine the cause, and pressing with his stick, found something hard, which, on a nearer inspection, proved to be a pot of gold. With part of this money he purchased the land, and settled in the parish; to which he bequeathed it on the conditions aforesaid.

"It was with pleasure," observes Mr. Taylor, in his "General Character of the Dog," "that I watched the motions of a grateful animal belonging to one of the workmen employed at Portsmouth dockyard. This man had a large cur dog, who regularly every day brought him his dinner upwards of a mile. When his wife had prepared the repast, she tied it up in a cloth, and put it in a hand-basket; then calling Trusty (for so he was properly named), desired him to be expeditious, and carry his master's dinner, and be sure not to stop by the way. The dog, who perfectly well understood his orders, immediately obeyed, by taking Page 471 the handle of the basket in his mouth, and began his journey. It was laughable to observe that, when tired by the way, he would very cautiously set the basket on the ground; but by no means would suffer any person to come near it. When he had sufficiently rested himself, he again took up his load, and proceeded forward until he came to the dock gates. Here he was frequently obliged to stop, and wait with patience until the porter, or some other person, opened the door. His joy was then visible to every one. His pace increased; and with wagging tail, expressive of his pleasure, he ran to his master with the refreshment. The caresses were then mutual; and after receiving his morsel as a recompense for his fidelity, he was ordered home with the empty basket and plates, which he carried back with the greatest precision, to the high diversion of all spectators."

Some years since, a distiller, who lived at Chelsea, in Middlesex, had a middle-sized brown cur dog, crossed with the spaniel, which had received so complete an education from the porter, that he was considered a very valuable acquisition. This porter used generally to carry out the liquors to the neighbouring customers in small casks, tied up in a coarse bag, or put in a barrow; and whenever the man thought proper to refresh himself (which was frequently the case), he would stop the barrow, and calling Basto (which was the dog's name), in a very peremptory manner bid him mind the bag;

and away he went to drink; and Page 472 frequently left the barrow in the middle of the street. Basto always rested near his trust, and sometimes apparently asleep; which induced many idle people, who, seeing a bag in the road without an owner, to attempt stealing the same. But no sooner had they endeavoured to decamp with the prize, than this vigilant creature flew at them with such outrage, as obliged them immediately to relinquish the undertaking; and glad were they to escape with a few bites and whole bones, and leave the tempting bait to catch other dishonest rogues, as it had done them.

One day, a person having particular business with the master, which required dispatch, went to the distillery adjoining the dwelling-house, thinking it very likely he might meet him there giving orders to the servant; and finding the outward door open, walked into the still-room: but no sooner had he gone a few steps than a fierce growl assailed his ears, and almost imperceptibly he was pinioned by fear to the wall. The affrighted person called loudly for help; but the family being at the other part of the house, his cries were fruitless. The generous animal, however, who had the frightened man close in custody, scorned to take a mean advantage of his situation by recommencing hostilities. He remained perfectly quiet, unless the delinquent attempted to stir—he then became as furious as ever; so that the prisoner prudently remained like a statue fixed against the wall, while Basto, like a sentinel on his post, kept a strict guard, lest he should Page 473 escape before the family arrived. In about twenty minutes the master, in coming from the parlour to the counting-house, beheld the prisoner, and Basto walking backwards and forwards beside him. The dog, by a thousand gesticulations, seemed to wish a proper explanation might take place. The master laughed heartily at the poor fellow's expense, as did he likewise when liberated; but he had ever after the prudence, when business brought him to the house, to ring loudly at the door, notwithstanding it frequently stood wide open.

A carrier on his way to Dumfries had occasion to leave his cart and horse upon the public road, under the protection of a passenger and his dog Trusty. Upon his return, he missed a led horse belonging to a gentleman in the neighbourhood, which he had tied to the end of a cart, and likewise one of the female passengers. On inquiry he was informed that, during his absence, the female, who had been anxious to try the mettle of the pony, had mounted it, and that the animal had set off at full speed. The carrier expressed much anxiety for the safety of the young woman, casting at the same time an expressive look at his dog. Trusty observed his master's eye, and aware of its meaning, instantly set off in pursuit of the pony, which he came up with soon after he had passed the first toll-bar on the Dalbeattie road; when he made a sudden spring, seized the bridle, and held the animal fast. Several people having observed the circumstance, and Page 474 the perilous situation of the girl, came to her relief. The dog, however, notwithstanding their repeated endeavours, would not quit his hold of the bridle; and the pony was actually led into the stable with the dog, till such time as the carrier should arrive. Upon the carrier entering the stable, Trusty wagged his tail in token of satisfaction, and immediately relinquished the bridle to his master.

A short time ago a large cur, belonging to a gentleman at Richmond, in Yorkshire, accidentally fell into a well, and for the moment he gave him up as lost. But as a sort of desperate effort to save the dog, he directed a boy to let down a rope he had into the well, in the hope that possibly it might catch around his leg or neck. No sooner, however, did the rope come within reach, than the dog seized it with his teeth, and the parties above finding it had secured him, began to draw up; when, about half-way up, he lost his hold and fell back. Again the rope was let down, and again the dog seized it, and he was drawn nearly to the mouth of the well; when his bite gave way, and the third time he fell into the water. Once more the rope was let down, and this time the dog took so thorough a hold, that he was brought triumphantly up; and when set down in safety, shook the water from his hair, and wagged his tail, apparently as proud of the exploit as the other parties were gratified with it.

Page 475

THE LURCHER.

This variety is smaller than the greyhound, with its limbs stronger and shorter, the head less acute, with short, erect, and half-pricked ears: the whole body and tail are covered with rough coarse hair; it is grizzly about the muzzle, of a pale sand-colour, or iron-grey, and of sullen aspect.

The lurcher is supposed to have been originally a cross between the greyhound and the shepherd's dog, re-crossed with the terrier; hence the quickness of his scent, his speed, and intelligence. The habits of this dog lead him to concealment and cunning, and he is seldom found in the possession of honourable sportsmen. He is often employed by poachers in killing hares and rabbits in the obscurity of night; and when taken to the warren, he lies squat, or steals out with the utmost precaution, and on seeing or scenting the rabbits, darts upon them with exceeding quickness or runs them down at a stretch, without barking or making the least noise. He is trained to bring the booty to his master, who often waits at some distance to receive it. One of these dogs will kill a great many rabbits in the course of a night. Col. Hamilton Smyth says, "The lurcher occasionally makes great havoc among sheep and deer, and acquires the wild scent of game. Sometimes these dogs become feral, when their owners happen to be captured and imprisoned. They have been regularly hunted with Page 476 hounds, but seldom destroyed, because when the chase came up with them, the pack seemed to be surprised at finding that it was only a dog they had followed. At other times, however, when a lurcher had snapped up, or attacked the game the pack was hunting, the dogs on coming up have torn him to pieces, as if he had been a wild beast."

Bewick says that in his time this breed was so destructive that it was proscribed, and is now almost extinct. "I have seen a dog and bitch of this kind," he observes, "in the possession of a man who had formerly used them for the purpose above described. He declared, that by their means he could procure in an evening as many rabbits as he could carry home."

"In the year 1809," says Capt. Brown, "I resided for some time on Holy Island, coast of Northumberland, and had occasion one day to be in Berwick at an early hour. I left the island on horseback at low-water, by moonlight. When I reached Goswick-warren, I came upon two men sitting by the side of a turf-dyke. I spoke to them; and while I was in the act of doing so, a dog of this breed approached with a rabbit in his mouth, which he laid down and scampered off.

Being convinced they were engaged in rabbit-stealing, I entered into conversation respecting the qualities of their dogs, which I was anxious to learn; and upon my declaring that I was a stranger, and that I would not divulge their delinquency, they readily gave me a Page 477 detail of them. They had scarcely commenced when another dog made his appearance with a rabbit, and laid it down, but did not, like his companion, make off when he had done so. One of the men said to him, 'Go off, sir,' when he immediately left them; and he told me he was a young dog, little more than a year old. They informed me, that such was the keenness of the older dog, and another which had shortly before died, for hunting rabbits and hares, that they would frequently go out of their own accord, when it was inconvenient for their owners to attend them, and that they invariably fetched in a hare or rabbit. Indeed, their ardour was such, that they would sometimes go to a rabbit-warren, at a distance of eight miles from their dwelling, in pursuit of game; in consequence of which it became necessary for their masters to chain them every night when they did not accompany them in this pursuit. The dogs never attempted to leave home during the day, for which reason they were allowed to go at full liberty. When the men intended on an evening to hunt rabbits, they threw down the sacks in which they carried their booty in a corner of their house, when the dogs lay down beside them, and would not stir till their masters took them up. These dogs scarcely ever barked, except on the way either to or from this plunder; on which occasions they always preceded their owners about fifty yards. If they met any person coming, they invariably made a noise, but never were known to bite any Page 478 one. I asked them if this was an instinctive property, and they informed me they were trained to it. As they found it necessary in various places to leave the highway to avoid villages, their dogs never failed to quit the road at the very places where they usually deviated, although at that distance before them. Sometimes one of the dogs would return back to the party while on the road, and wag his tail, but they seldom or never did so together; and if he showed a desire to remain by his master, the latter had only to say, 'Go on, sir,' when he set off at full speed to his post as one of the advanced guard. During the time I was conversing with them these dogs brought in seven rabbits."

The following curious relation, in which a lurcher signalised himself characteristically but fatally, we had from a sporting clergyman of one of the midland counties. A gentleman kept a pack of five-and-twenty couple of good hounds, among which were some of the highest-bred modern foxhounds, and some as near to the old bloodhound as could be procured. They were high-fed and underworked; of course, somewhat riotous. One day, after a sharp run of considerable length, in which the whole field, huntsman, whipper-in, and all, were suddenly thrown out, Reynard, in running up a hedgerow, was espied by a lurcher, accompanying the farmer his master. The dog instantly ran at the chase; and being fresh, chopped upon it as he would have done upon a rabbit or hare. The fox turned and fought Page 479 bravely; and whilst the farmer was contemplating with astonishment this singular combat, he was destined to behold a spectacle still more remarkable. The hounds arrived in full cry, and with indiscriminate fury tore both the combatants to pieces; the whipper-in, and the proprietor of the pack, and two or three gentlemen the best mounted, arriving in time to whip the dogs off, obtain the brush, and pick up some scattered remnants of the limbs and carcase of the poor lurcher.

THE BAN DOG.

This variety, which seems almost extinct, is lighter, smaller, and more active than the mastiff, from which he is descended by a cross with the foxhound. He is not nearly so powerful a dog as the former, but is more fierce in his natural disposition; and from his descent possesses a finer sense of smelling. His hair is rougher, generally of a yellowish or sandy grey, streaked with shades of black, or brown, and semi-curled over his whole body, excepting his legs, which are smooth. Although he generally attacks his adversary in front, like the mastiff and bull-dog, it is not his invariable practice, for, he is sometimes seen to seize cattle by the flank. His bite, says Bewick, is keen and dangerous.

Page 480 Two near neighbours in the county of Suffolk, a tanner and a farmer, entertained great friendship for each other, and kept up a close intimacy by frequent visits. The tanner had a large ban-dog for watching his yard, which, from some unknown cause, had conceived such an inveterate hatred to the farmer, that he could not go with safety to call on his friend when the dog was loose, and on this account the tanner loaded him with a heavy clog, that he might not be able to fly at him.

As the farmer and one of his ploughmen were going about the grounds together one day, the latter espied at a distance something on a stile. As they drew near, they perceived it was the tanner's dog, which, in attempting to leap the wall, had left the clog on the other side, and was thereby almost strangled. The ploughman, knowing the enmity which the dog had to his master, proposed to despatch him by knocking him on the head; but the latter was unwilling to kill a creature which he knew was useful to his friend. Instead of doing so, he disengaged the poor beast, laid him down on the grass, watched till he saw him recover so completely as to be able to get up on his legs, and then pursued his walk. When the farmer returned to the stile, he saw the dog standing by it, quite recovered, and expected an attack; but, to his great astonishment, the creature fawned upon him, and expressed his gratitude in Page 481 the most lively manner; and from that time to the day of his death he attached himself to his benefactor, and never could be prevailed upon to go back to his former master.

Page 482

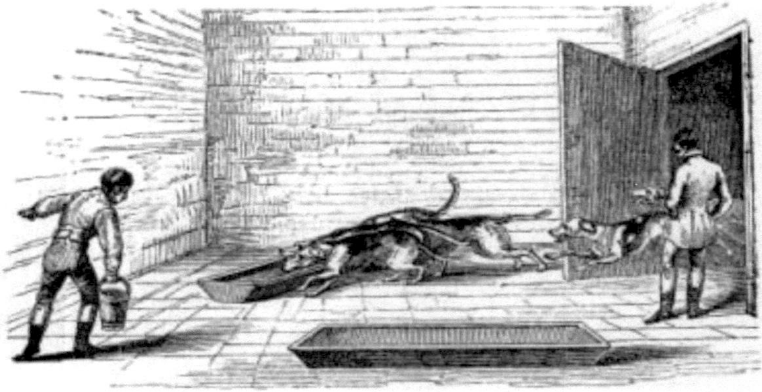

FEEDING HOUNDS.

ON THE FEEDING AND MANAGEMENT OF DOGS.

Gathered from various authorities by H. G. Bohn.

A few words may not be out of place here on the feeding and management of dogs. For all else which concerns Canine Science the reader cannot do better than consult, among modern works, "Youatt on the Dog," "Blaine's Canine Pathology," the article "Dog" in the Encyclopædia Britannica or Penny Cyclopædia, "Hutchinson on Dog-Breaking," "Radcliffe on Fox-Hunting," "Mayhew on the Dog," or, "Colonel Hamilton Smith on Dogs," forming two of the vols. of Jardine's Naturalists' Library.

The natural food of the dog is flesh, and it is found that those in a wild state prefer it to every other kind of nutriment, but as raw meat engenders ferocity, it should not be given too freely, especially to house-dogs and such as are not actively exercised. The dog can subsist on many kinds of food, and it is a curious fact, that when fed entirely on flesh he will sometimes get lean; because, as has been well observed, it is not on what animals eat that they thrive, but on what they digest. The diet of sporting dogs in full work should, it is said by some, consist of at least two-thirds of flesh, with a judicious mixture of farinaceous vegetables; but there is great diversity of opinion on this subject, and in France they are fed almost exclusively on soaked bread. Dogs, it is generally said, should have free access to fresh water, and the pans be cleaned out Page 483 daily; but some feeders, we are told, and it seems strange, limit the supply of water, and substitute moistened food. A piece of rock brimstone kept in the pan will be found useful.

Although the dog is naturally a voracious animal, he can endure hunger for a very great length of time, and be brought by habit to subsist on a very scanty meal. In the Memoirs of the Academy of Sciences it is stated, that a bitch which was forgotten in a country-house, where she had access to no other nourishment, lived forty days on the wool of an old mattress which she had torn to pieces and digested.

An extraordinary instance of a similar kind occurred with a terrier bitch, named Gipsy. One day, when following her master through a grass-park near Gilmerton, it happened that she started a hare. During the pursuit her master suddenly lost sight of her, and in a few days she was considered either killed or lost. Six weeks afterwards a person happening to look down an old coal-pit, was surprised to hear a dog howling. He lost no time in returning to the village, and having procured a hand-basket, let it down by a rope into the shaft; the dog immediately leapt into it, and on being brought to the surface, proved to be Gipsy, worn to perfect skin and bone. How she had existed in this subterranean abode, and what she had found to support her there, it is impossible to tell.

Stag-hounds, fox-hounds, harriers, and beagles, are generally fed on oatmeal,—some add well-boiled flesh to it once in two days,—and the older the meal is the better. Store sufficient for twelve or eighteen months' consumption ought, therefore, always to be kept by those who have a pack; and before used should be well dried, and broken into grits, but not too fine. It is best kept in bins in a granary, well trodden down. Some persons are in the habit of using barleymeal unprepared, but this is thought by many to be less nutritious. Others are of opinion that oatmeal and barleymeal in equal proportions form a preferable food. In either case the meal should be made into porridge, with the addition of a little milk, and occasionally the kitchen offal, such as remnants of butchers' meat, broth, and soups, the raspings and refuse of bakers' shops, or hard, coarse, sea-biscuit (sold as dog-biscuit), well soaked and boiled with bullocks' liver or horseflesh.

Well-boiled greens—or mangel-wurzel boiled to a jelly—are an excellent addition to the food of all dogs, and may be given twice a-week; but they ought to be discontinued during the shooting-season with pointers, setters, cockers, and greyhounds; and also during the hunting season with foxhounds, harriers, and beagles, as they are apt to render the bowels too open for hard work.

Page 484 Flesh for dogs should be first thoroughly boiled and then taken out before the oatmeal is added to the broth, and left to cool. Indeed, some feeders think that the food of a dog should always be perfectly cold. At any rate, care must be taken not to serve it out "too hot," although, in general, dogs are sagacious enough not to scald themselves, as we see in Landseer's exquisite little picture on the opposite page.

Dogs which are hard worked are by some said to be the better for having two meals a-day—a very light one of mixed food in the morning before going out, and a full meal, principally of flesh, on their return in the evening; but, as a general rule, one good meal a day, towards the evening, is sufficient, and they may be left to pick up what they can: indeed the dealers never give more than one meal a-day. Bones to pick may be allowed them occasionally, but hard bones in excess are likely to wear and damage the teeth. Nothing is better than paunch, tripe, or good wholesome horse or cow-flesh, boiled, and the liquor mixed well with oatmeal porridge; the quantity of each about equal. If horse or cow-flesh is not to be had, graves, in moderate quantity and well scalded, are a tolerable, though not very desirable, substitute. They are generally broken small, mixed with about one-half the quantity of oatmeal, then thoroughly soaked in boiling water, and well stirred; or, a better way still is to boil them together like porridge.

Dogs, like men, require a change of food, and it has been strongly asserted that barleymeal and oatmeal, without change, predisposes to cutaneous disease, and even produces it; therefore, a judicious feeder, like a good cook, will contrive to vary his bill of fare. Porridge and milk, dog-biscuit, farinaceous food, the scraps of the kitchen, the offal of bullocks or sheep, which should be well boiled, make an excellent variety;—but we would by no means recommend too frequent a repetition of the latter food. Potatoes are also good, and although not so nutritious, or easy of digestion, as oatmeal, are less heating.

Care should be taken never to present more to a dog than he will eat with a good appetite; and when oatmeal and barleymeal are given mixed, the former should first be boiled for twenty minutes, and then the latter added, and boiled only for about eight or ten minutes. This meal should, however, never be given in the hunting season, as it is too heating, and occasions the dogs to be perpetually drinking. Their food ought, as a general rule, to be given to them pretty thick, as thin porridge does not stay the stomach so well. The feeding-troughs for hounds should be sufficiently wide at the bottom Page 485 and carefully cleaned out and scalded with hot water every second day.

During the hunting season hounds should have sulphur mixed up with their mess once a-week, in the proportion of 3 drachms to each. At the end of the season the same quantity of sulphur should be given, with the addition of 1½ drachms of antimony. After a hard day's work a meal of horse-flesh may be given them, as fresh-killed as possible, or bullocks' paunches or sheeps' trotters, all of which should be well boiled.

Greyhounds should be fed principally on animal food, such as sheeps' trotters or neats' feet, boiled or stewed down and mixed with bread, and given moderately in the morning and afternoon, (the dog never being allowed on any occasion to eat a great quantity at once,) or on other hand meat, as it will enlarge and strengthen the muscular fibre without increasing the cellular tissue and adipose substance, which has an invariable tendency to affect their breathing. The butchers' meat should be of the best quality, and not over-fat, as greasy substances of all kinds are apt to render the body gross and the skin diseased. After they have been coursed they should be well brushed, a little oil being used in the operation.

The kennels of greyhounds should be kept comfortably warm and dry, be frequently replenished with dry and clean straw, and properly ventilated. Indeed, nothing is more essential to the health and efficiency of all dogs than pure air and cleanliness. Their beds should, if possible, be placed on a wooden bench, or at least on some dry position. On attention to cleanliness depends, in some degree, the dog's exquisite sense of smelling; for, if accustomed to strong or disagreeable effluvia, he will be but ill-adapted to trace the fall of a deer, or scent of a fox. Indeed, even animal food too freely given is said to have a prejudicial effect upon the nose of a sporting dog.

A dog employed in watching premises should not be needlessly exposed to the damp or cutting night winds; but placed in as dry and sheltered a situation as possible. If kept in the dwelling-house he should have a place appropriated to his night's rest; this may be an open box, or a basket, with a piece of carpet or blanket, or clean straw at the bottom: if either of the former it should be often beaten, to free it from fleas or nits, which soon infest it, and frequently washed and dried.

Damp is exceedingly injurious to dogs, and is very likely to produce diseased lungs, rheumatism, and lameness in the shoulder and limbs.

Page 486 To the preceding instructions, for which the compiler is chiefly indebted to the works of Capt. Thomas Brown, Youatt, and Blaine, and to the practical information obtained from Mr. Herring of the New Road, and Mr. William George, an extensive dog-fancier at Kensall New Town, may be appropriately subjoined a lively chapter from the recent work of Mr. Francis Butler, a leading American authority on the subject.

"It is more important to understand the management of a dog, than to be possessed of a thousand nominal remedies for the cure of his various ailments; inasmuch as the antidote is at all times preferable to the cure.

"I shall first throw out a few hints on the Management of Pets. Whilst many are sacrificed for lack of necessary attendance, there are thousands who perish prematurely from overdoses of kindness. Delicate breeds of dogs certainly require great care and attention in rearing; but overstrained tenderness is often more dangerous than culpable neglect. The dear little creature that is allowed to lay under the stove, is stuffed with delicacies two or three times a-day, and is never allowed to breathe the fresh air, except under a cloudless sky, is more subject to colds, fits, rheumatism, sore eyes and ears, worms, &c., than the worthless mongrel which was raised on the street, neglected and despised. The tenderly-nursed pet is affected by every change of atmosphere, and subjected to a variety of diseases unknown to the dog that has been hardened from his birth. I ask you, then, neither to stuff nor starve; neither to chill nor burn.

"A house-pet should always have a sleeping-place allotted to him, warm and comfortable, not near the fire, nor in the damp. Anything round is best for an animal to lay in; such as a tastefully ornamented box. In cold weather it should not be larger than to contain him comfortably. It is best for the following reasons: he may keep himself perfectly warm, and his bed may be made exactly to fit him; it also takes up less available space than any other shape. He should never be fed to the full; neither excited to eat when he appears disinclined. Lack of appetite, so common to pampered favourites, is generally the result of an overloaded stomach and disordered digestion. This is easily cured by medicine, but more safely and simply without it. Fast him for twenty-four hours; after which, keep him on half his ordinary allowance. If this agrees with him, and he keeps in fair condition, continue the regimen.

"Nursing in the lap is injurious; not in itself, but the animal is thereby subjected to constant chills, in emerging Page 487 from a snoozy warmth to a cold carpet or chilly bed. A dog accustomed to the lap is always shivering after it, and renders himself quite troublesome by his importunate addresses. A moderate share of nursing is well enough, but should be indulged in only as an occasional treat. Great care should be taken in the washing of delicate dogs. When this operation is performed, they should be rubbed perfectly dry; after which they should be covered, and remain so till the shivering has completely subsided. [T] The water should be only blood-warm; it is far better than hot, and not so likely to give the animal cold. Injudicious washing and bad drying are productive of running sore eyes, more especially visible in white poodles, where the hair is long and woolly, retaining the moisture.

"Once a fortnight is often enough to wash any dog but a white one. Washing has very little effect in the destruction of vermin. Fleas can live some time under water; which I have often thought only makes them bite the harder and stick the closer, when reanimated from their temporary torpidity. If 'Butler's Mange Liniment and Flea Exterminator' cannot be obtained, the animal may be well sodden with soft soap and washed about ten minutes after. This cannot be done with safety, except in warm weather. In cold weather, the comb may be used immediately after the application of the soap, as the fleas will then be too stupid to effect their escape. 'Butler's Liniment' destroys all vermin instantaneously, without risk of injuring the animal; and the quadruped may be rinsed one minute after. No flea will remain alive; the skin will be thoroughly cleansed, and the coat beautified. Dogs should never be allowed to suffer the torment imposed on them by these detestable vermin. If the owners could only realise the importance of ridding them of these ever-noisome pests, there would be far less of snappishness, mange, fits, &c. I have seen animals literally worried to death by fleas, perfectly exhausted from incessant irritation, at last worn to a skeleton, and gradually extinguished by a creeping consumption. Besides, who (for his own personal comfort), would not rid his immediate vicinity of a worthless mob of blood-suckers awaiting the first favourable opportunity of regaling themselves on human blood? If your dog lie on straw, burn it once a week, as fleas harbour and propagate in the tubes of the straw. If the bed be carpet, or anything similar, let it be often cleansed Page 488 or changed. Vermin revel in filth, and their extirpation depends mainly on cleanliness.

"By attending to the general health of a dog, much disease may be avoided; indeed, this is far more essential than prescriptions for a cure. It is very easy to carry off a slight indisposition by gentle purgatives and a reformed diet: whilst confirmed disease is often difficult to combat, as few of the canine race can have the advantages which are ofttimes essential to their restoration. The eyes, the nose, the gums, the hair, the breath, should be carefully noted. The eyes may be red or pale, sunken or protruded; the nose may be hot, or dry, or matted with dirt; the gums may be pale, &c. It will require but little experience to discover a disorganisation, which may be easily detected by him who has noticed the healthful appearance of the different parts and their variation under indisposition.

"If you are in the habit of keeping your dog on the chain, let him at least run a few minutes every day. If he be kept indoors, he should also be allowed a little daily exercise outside. Change of air [U] and diet will sometimes renovate when all remedies fail: a change from city to country, from greasy meat to fresh milk, from a confined yard to the green fields, will generally recruit him without the aid of medicine. Nature (to whom physicians are so deeply indebted for so many wonderful restorations), often effects a cure unaided, which might have defied the efforts of Apothecaries' Hall.

"In summer, particularly, be careful to provide a supply of fresh water and a cool shelter from the sun. Never take your dog out during the intense heat of the day; this is very apt to produce fits, often resulting in sudden death. Early in the morning is preferable for summer exercise.

"The kennel should be located in a shady spot during the summer; in winter it should be sheltered from the wind, and so placed as to enable the dog to enjoy the sunshine at will. Above all things, never chain a dog where he cannot screen himself from the sun's rays. He must have the option of sunshine or shade. He should not be allowed to drink water that has been standing in the sun, or is otherwise damaged. If you should chance to forget to feed him for forty-eight hours, he would not run as much risk of injury, as during three hours of thirst in hot weather. There should be a piece Page 489 of joist under each end of the dog-house, to keep it off the ground, in order to avoid dampness. In summer an excavation, two or three feet in depth, should be made under it, and left open at both ends, that the animal may have a cool retreat during the heat. Those who do not object to a trifling expense, may have the house posted on a large paving-stone, with an excavation under it, as before recommended. All burrowing animals seek the earth in hot weather. Everything on the surface is heated; their own instinct dictates the most reasonable method of sheltering themselves from the heat, at the same time absorbing the cool exhalations from the ground. In southern climates, especially, this method is all important. In this manner I have kept dogs from the polar regions, in comparative comfort, whilst many native-born and neglected have been scalded into fits, paralysis, rabies, or hydrophobia.

"In the hot season, with young dogs, raw meat should be avoided, except it be quite fresh, and then they should not be over-fed, especially if debarred of abundant exercise, and excluded from their own natural medicine, grass. A dog will often thrive better on raw meat than on any other food, and will grow larger; but he should be fed with discretion, and his health attended to, should his diet visibly disagree with him. [V] He will grow fatter and be more healthy on moderate meals than if overgorged. The better plan is to ascertain his average consumption, and then allow him a little less. Keep his digestion in good order, and disease will rarely trouble him. His coat and ribs will generally indicate whether he be sufficiently cared for, whether he be sick or sound in his digestive organs; feed him always in the same place, and at the same hour: once a day is sufficient, if he be over six months old. By being fed only once a day he is less choice, and will consume what he might refuse, if his appetite were dulled by a previous meal.

"Should you require your dog to be watchful at night, feed him in the morning; if you would have him quiet at night, feed him late, and don't leave him bones to gnaw. Dogs are pretty quiet, during the digestive process, when left to themselves, and should not have much exercise after a heavy meal. They should only be lightly fed before training-lessons, or on sporting days; on the latter occasions a little refreshment may be administered as occasion may require. Those Page 490 kept in-doors should be allowed to run a little after meals, when they generally require an evacuation.

"If a dog be regularly exercised he will seldom even soil around his kennel, and a healthy house pet is rarely troublesome, except after eating. If a dog be uncleanly in the house, he should decidedly be broken of it, although it would be useless to correct him unless he has a fair opportunity of avoiding it. He should be invariably taken to the spot, be sufficiently twigged there, and unceremoniously scolded into the yard. The punishment will be far more justly administered if the animal be let out at regular intervals; this being done he will not attempt to infringe the law, except in cases of dire necessity.

"I am satisfied as a general rule, that a well-amalgamated mixture of animal and vegetable is the most healthful diet for dogs of all ages, breeds, and conditions. Dogs living in the house should on no account be fed on raw meat, as it gives them a very offensive smell, and is in other respects very unsuitable."

FOOTNOTES:

[A] Daniel's "Rural Sports."

[B] Daniel's "Rural Sports."

[C] Thornton's "Instincts."

[D] "Sportsman's Cabinet."

[E] Ballet, in his "Dissertations sur la Mythologie Française," shows that this popular story of the dog of Montargis is much older than the time of Charles V.; and that Albericus, an old monkish chronicler, records it as happening in the reign of Charlemagne, anno 780.

[F] See the entire poem in Tomkins' "Beauties of English Poetry." 18mo. 1847.

[G] "I fear this is a sad geological anachronism; however, I cannot but hope that the Irish wolf-dog will yet be found in some cavern, associated with the prototypes of Ireland's earliest heroes who peopled the land soon after it emerged from the deep,

'Great, glorious, and free,
First flower of the earth and first gem of the sea.'"

[H] O'Keefe, "Wicklow Gold Mines."

[I] A similar instance of canine intelligence will be found in p. 51 of the present volume.

[J] "The Sportsman's Cabinet."

[K] Tenbeia portus est Cambriæ meridionalis, ubi Belgarum colonis a rege, ut fertur, Henrico primo locata est. Horum posteri a circumjacente Celticæ originis populo lingua etiam nunc omnino discrepant.

[L] Infinitivo, quem vocant, hoc in ier desinente solus credo, inter, melioris notæ, quos habemus, elegorum scriptores usus est Catullus: sed qualis ille Poeta! sed quantus in omni genere Latini carminis et artifex elegantiæ et magister!

[M] His master's pocket-book, with which Tippo, the only living creature saved from the wreck, came ashore.

[N] See Bewick's "Quadrupeds," p. 306, 1st ed.

[O] A celebrated portrait painter, and Secretary to the Scottish Academy of Painting. This gentleman also excelled in the portraits of animals.

[P] "Sometimes the members or domestics of the convent have been sufferers in their efforts to save others. On the 17th of December, 1825, three domestics of the convent with two dogs descended to the vacherie, on the Piedmontese side of the mountain, and were returning with a traveller, when an avalanche overwhelmed them. All perished except one of the dogs, which escaped by its prodigious strength, after having been thrown over and over. Of the poor victims, none were found until the snow of the avalanche had melted in the returning summer, when the first was discovered on the 4th of June, and the last on the 7th of July."

[Q] Mrs. Grosvenor, now of Richmond, Surrey.

[R] For other instances of speaking dogs see *ante*, p. 49.

[S] In p. 147 a similar anecdote has been recorded of a Newfoundland dog and a spaniel; and in p. 221 an instance is given of the revenge taken by a Colley on a tailor's dog.

[T] Or if the weather be fine and warm they may run out and dry themselves.—Ed.

[U] Sea-air, however, especially during long sea-voyages, perhaps in connexion with salt meat, has been known to produce the distemper in dogs.—Ed.

[V] House-dogs fed on raw meat, bones, and liver, soon become offensive neighbours; the more so in proportion to their want of outdoor exercise.—Ed.

Printed by G. Barclay, Castle St. Leicester Sq.